AF863765

ASTROLOGICAL CRIME CHARTS

ABOUT THE AUTHOR

Arlene DeAngelus has been involved in astrology for more than twenty-five years and has practiced as a professional astrologer since 1980. She has a Certificate of Merit from Ivy Goldstein-Jacobson and a Certificate of Proficiency from The Mayo School of Astrology in England.

She was a columnist and wrote the Astrological Forecast in the Psychic Guide/Body Mind Soul magazine from 1984 to 1988. She has served as faculty in 1986 through 1998 and will again in 2000 at the American Federation of Astrologers Conventions lecturing on "Comparing Murder/Suicide Charts," "Looking for Clues in Missing Children's Charts," and "Crime Charts." DeAngelus has also lectured for the National Council for GeoCosmic Research, Inc., Central NY Astrological Association and other organizations, high schools and colleges.

Arlene's academic background includes a BA in Communication Arts from The College of Saint Rose in Albany, New York where she is currently enrolled in graduate studies. She lives in Clifton Park, New York with her husband, Ron, and their Bengal cat, Kymba.

TO WRITE TO THE AUTHOR

The current status for all cases in Crime Charts is given as of the writing of this book. The author welcomes your comments and questions, but cannot guarantee that every letter written received will be an answered. Where the cases are still open, your comments that are directly related to the case itself, will be given to the proper authorities.

The author also participates in seminars and workshops. To contact the author, write to:

Arlene L. DeAngelus
18 Montgomery Way
Clifton Park, New York 12065
E-mail: ADeAngelus@AOL.com

Please enclose a self-addressed, stamped envelope for a reply.

ASTROLOGICAL CRIME CHARTS

A "What To" and "How To" of Crime Charts

Arlene DeAngelus

Dedication

This book is dedicated to my husband,
Ronald DeAngelus,
and to my teacher,
Ivy Goldstein-Jacobson

TABLE OF CONTENTS

CHARTS

ILLUSTRATIONS

PICTURES

MAPS

FIGURES

INTRODUCTION

At sometime every astrologer is asked by a friend or acquaintance to help the family, or the investigators, when someone has been murdered, disappears or maybe committed suicide. It can also be only a personal interest in a particular crime that has taken place. Then, further the question often arises as to whether the use of astrology can be successfully utilized as a tool in criminology. Can one's astrological chart give any additional clues which are needed to aid investigators in the solving a crime? *Crime Charts* is a result of those questions and tells "what to" and "how to."

We know that astrology is the study of the cycles of the heavenly bodies and that these cycles affect human affairs. Astrologers are able, therefore, to see the possibility of one's charts providing additional clues when a crime has been committed. In order for an astrologer to prepare an astrological birth chart, or natal chart, the birth date, the birth place, and an accurate birth time are required. This chart can then be used with the transiting, progressed and solar planets to show the current trends at any given period in one's life.

Crime Charts gives you the background of actual victims in several crime cases as well as the calculations of accurate natal, progressed, transit and solar charts. These charts are examined through the eyes of an astrologer for ways in which clues can be found in the crime victim's charts. Yods, critical degrees, the 29 and 00 degrees of a sign and eclipses play an important part in a crime victim's chart and will be reviewed to see if they can also offer additional clues in a criminal case.

It should be remembered that astrology is an interpretative science which is dependent entirely upon the astrological knowledge and interpretative capabilities of the practitioner. Although, astrological placements and aspects will be noted, it is essential for interpretation purposes that one know the basic meanings of the astrological signs, planets and houses.

CRIME CHARTS

Recent reports issued by the Federal Bureau of Investigation say that the nation's violent crime rate has fallen for several straight years. Murders and robberies show the sharpest declines and have dropped by about nine percent. It further states that in the last three years, the murder rate has plummeted 25 percent.

Although violent crime statistics have been dropping in large cities, suburban counties and most mid-size cities, it has risen in rural counties. New York City's violent crime statistics have also shown a drop and are at their lowest rate since 1968. Several large cities report the same trend, while other cities such as New Orleans have shown an increase in the percentage of violent crime. The decreasing crime rates in some areas are encouraging, but violent crime is still a plague in our country and many cities have alarming statistics.

Statistics show that there are still slightly under 1,000 homicides in New York City alone each year even with the current decreases. According to a 1997 report in *The Washington Post* more than 1.3 million people in the United States suffer injuries caused by violent attacks annually. Additionally, thousands of children are abducted across the country each year.

There is a possibility that every person will experience violent crime either personally or through a family member or a friend at some point in his or her life. Many of these violent crimes are solved quickly, while others may take several months or even years. Other crimes are never solved.

As an astrologer, you may experience crime vicariously through a request from one of your clients, or you may read a newspaper article about the controversial death of a prominent person which piques your interest. One such death that comes to mind is that of Vincent Foster. Reading a newspaper account of Foster's death you might wonder, "What really happened here? Was it a suicide or a murder?" Once you have obtained the person's birth information, you run a chart and sit down to examine it. What in that chart might answer some of the questions you have regarding the circumstances surrounding this death?

Sometimes, a cursory look at the aspects in the chart can give an indication as to whether the death was a suicide or a murder. Aspects that are indicative of emotional depression, mental instability and general unhappiness may suggest a pending suicide. However, the possibility of suicide usually is seen in the chart months before it actually occurs. On the other hand, aspects that indicate aggressiveness, competitiveness and combativeness can often indicate the possibility of a murder. It must also be remembered that these aspects occur repeatedly in charts, and they do not mean that everyone will commit suicide or be murdered. The chart needs to be examined in more detail to find additional corroborative data.

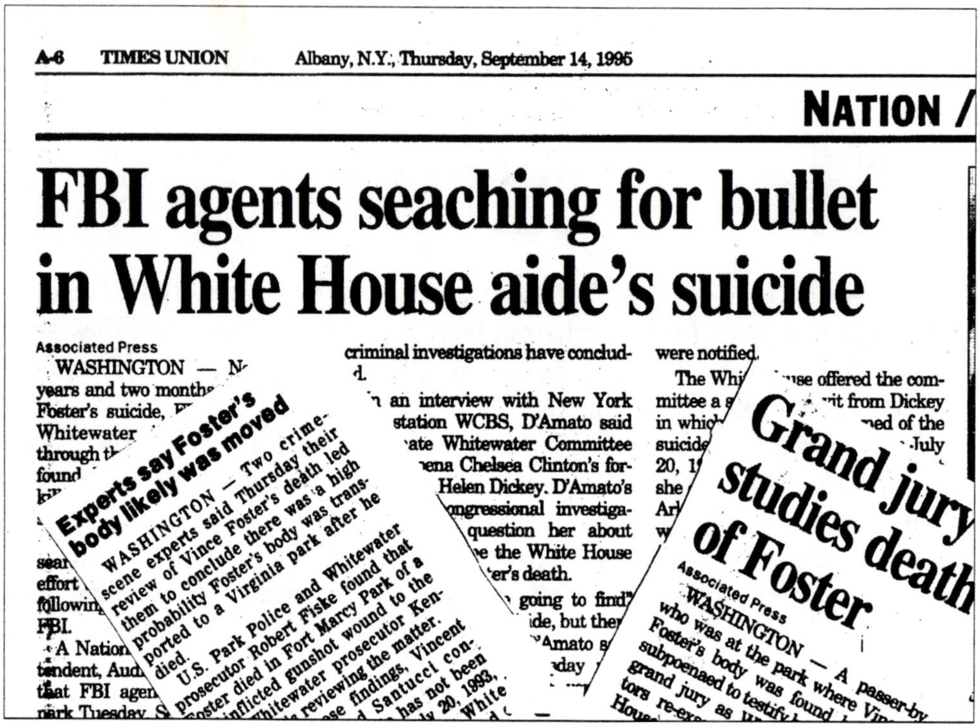
A-6 TIMES UNION Albany, N.Y., Thursday, September 14, 1995

NATION

FBI agents seaching for bullet in White House aide's suicide

Experts say Foster's body likely was moved

Grand jury studies death of Foster

Illustration 1 - Albany Times Union (Foster)

Where Do You Begin?

A basic knowledge of the houses, signs and planets enables the astrologer to review and interpret a crime chart. This book focuses on the similarities found in the natal, progressed, transit and solar return charts of crime victims. Although the charts are calculated in the Tropical Placidus system using accurate times of birth, use the calculation system that works best for you.

Begin by noting the house placements, degrees and aspects of the natal, transiting and progressed planets, the progressed Midheaven and the progressed Ascendant. Is any placement changing signs, is it stationary, retrograde or direct, or at 29 or 00 degrees? If so, those placements should be given extra attention.

Are any of the planets, the Midheaven or the Ascendant at a Critical Degree?

Cardinal (Ari, Can, Lib, Cap)	0	13	26
Fixed (Tau, Leo, Sco, Aqu)	9	21	
Mutable (Gem, Vir, Sag, Pis)	4	17	

Figure 1 - Critical Degrees

Are any transiting outer planets forming major aspects to the natal chart? Note the aspects allowing an orb of approximately two (2) degrees. The major aspects used in this book are:

Conjunction	0	Trine	120
Sextile	60	Inconjunct	150
Square	90	Opposition	180

Figure 2 - Astrological Aspects

Each aspect produces a particular energy in its connection to the planets:

A **Conjunction** (con) is two planets at the same degree. This aspect gives Emphasis, Cooperation and Assistance.

A **Sextile** (sxt) is two planets that are separated by 60 degrees. This aspect gives Opportunity, Understanding and Communication.

A **Square** (sqr) is two planets that are separated by 90 degrees. This aspect gives: Challenges, Obstacles, Conflict and Stress.

A **Trine** (tri) is two planets that are separated by 120 degrees. This aspect gives Ease, Harmony and Luck.

A **Quincunx** or **Inconjunct** (inc) is two planets separated by 150 degrees. This aspect gives Adjustment, Reorganization and Disharmony.

An **Opposition** (opp) is two planets that are separated by 180 degrees and are opposite of each other. This aspect gives: Awareness, Compromise and sometimes Opposing Factors.

A planet or the angles reach a point of culmination at the 29th degree of any sign. Both 29 and 00 degrees should be noted when interpreting the natal, transit, progressed or solar return charts.

29 degrees	A climactic or turning point which brings the matter in question to a climax.
00 degrees	A new course of action has been set, either through one's choice or by force.

Ivy M. Goldstein Jacobson states in *Simplified Horary Astrology* "...if any planet or cusp is in 29 degrees at birth the Moon will progress to aspect it and cause a disturbance which upsets the status quo there, especially if any change is under way at the time." (1) Jacobson further says, "The 29th degree shows some misfortune connected with the matter; the person or matter asked about is changing, at the end of his rope or patience, or desperate." (2)

The planets rotate and form major aspects in the following approximate cycles:

Cycle	Square	Oppose	Square
Jupiter 12 Years	3	6	9
Saturn 29-1/2 Years	7.5	15	22.5
Uranus 84 Years	21	42	63
Neptune 165 Years	41-42	83	

Figure 3 - The Planet's Cycles

The degrees of an event such as the Solar and Lunar Eclipses should also be noted. A chart is activated by a Lunar Eclipse for a period of six (6) months and a Solar Eclipse for a period of one (1) year.

Note the aspects that form a yod. The aspects forming a yod are two inconjuncts joined by one sextile. The yod is sometimes referred to as the "Fingers of Fate," "Destiny Mission" or the "Special Life Purpose. "A yod shows a special reason or purpose chosen by the soul to accomplish, to fulfill or to complete.

This book is not a book of interpretations, but instead it will focus on the aspects and placements that need to be interpreted when reading a crime chart. The well-known texts on interpretation, keywords, etc. referenced in delineation of this book will be mentioned for your information and possible use. Complete chart information is included so charts can be reconstructed without a computer.

Lastly, look at the charts for signatures of the events which have occurred. Become familiar with the basic keywords for the signs, planets and houses. An excellent reference source for these keywords is *The Rulership Book* by Rex E. Bills, and these keywords will be used with the yods as examples in this book.

Other General Information

The charts used in this book are the natal, transit, progressed and solar. They have been produced using Matrix Win Star. The progressed charts are secondary progressions with a solar arc Midheaven and are prepared for the birth place. Transit charts have also been prepared for the birth place, and solar return charts have been relocated to the place of residence for the year in question.

The placements for the progressed and transit charts are listed in their natal house positions while the planets for the solar return charts are listed in their solar house positions. Retrograde placements have been marked with an R.

Theodore Bundy's Progressed Chart for 1988

Theodore Bundy was a precipitator of crime, and in early 1989 he was executed for those crimes. His progressed chart for 1988 is a very good example of many of the mentioned placements. Bundy's progressed Midheaven is at 29:12 Gemini and his progressed Ascendant is at 29:20 Virgo. Progressed Jupiter, ruler of the law, is at the critical degree of 21:15 Scorpio in the fourth house ruling end of life.

Bundy's farewell

'Thank God, it's finally over'

By George Tibbits
Associated Press

SEATTLE — Ted Bundy's execution Tuesday ended an ordeal for some families of his victims, while others found it hard to comprehend that a moment they had sought for years had finally arrived.

"Good," sighed Vivian Rancourt, mother of Bundy victim Susan Rancourt, when reached at her LaConner

Serial killer dead at dawn

Combined wire services

STARKE, Fla. — The life of Theodore Robert Bundy came to an end as the sun rose Tuesday. The serial killer *linked by the FBI to dozens of murders* nationwide was electrocuted at Florida State Prison.

Prison doctor Frank Kilgo pronounced Bundy dead at 7:16 a.m.

It was the final page of a violent saga that gripped America for more

THEODORE BUNDY
'I don't want to die'

Illustration 2 - Schenectady Daily Gazette, January 25, 1989 (Bundy)

The progressed Moon always brings change, and we find the Moon in the natal eleventh house, indicating a change of objectives. The progressed Moon is at the critical degree of 00:06 Cancer, in its own sign, and the natural ruler of the fourth house. Matters of the fourth house rule the end of life or the manner of one's death.

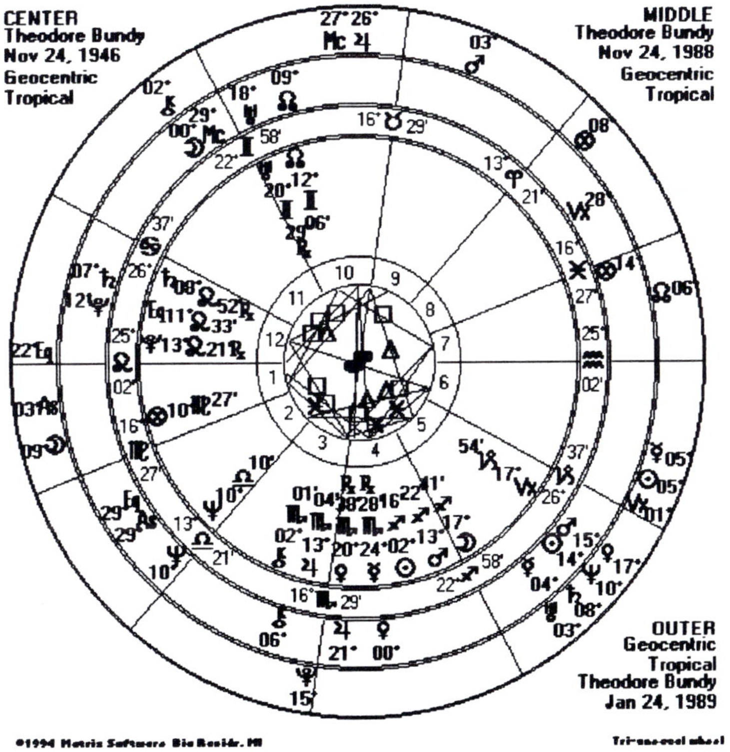

Chart 1 - Theodore Bundy's Natal, Progressed Chart for 1988
and Transits for January 24, 1989

Bundy's progressed Venus in the natal fourth house at 00:21 Sagittarius inconjuncts the progressed Moon at 00:06 Cancer. Appropriately, Venus rules justice and Sagittarius rules law, courts and long journeys. On January 24, 1989, Ted Bundy died in Florida's electric chair for the 1978 murder of twelve-year-old Kimberly Leach of Lake City, FL. His electrocution brought the matter of justice to the 29 degree climactic end.

Theodore Bundy's Solar Return Chart (Starke, FL)

In Bundy's Solar Return Chart for November 24, 1988, solar Venus, ruling sexual energy, is found at 00:57 Scorpio in the first house of self. Venus inconjuncts retrograde solar Jupiter at 00:53 Gemini in the eighth house. Solar Uranus is at 29:32 Sagittarius in the solar third house. The Sagittarian court of law as indicated by the placement of this Solar Uranus at 29 degrees and its ruling of death used the electricity ruled by this planet for Bundy's execution. Theodore Bundy's execution was the culmination or turning point of his inappropriate actions and sexual behavior.

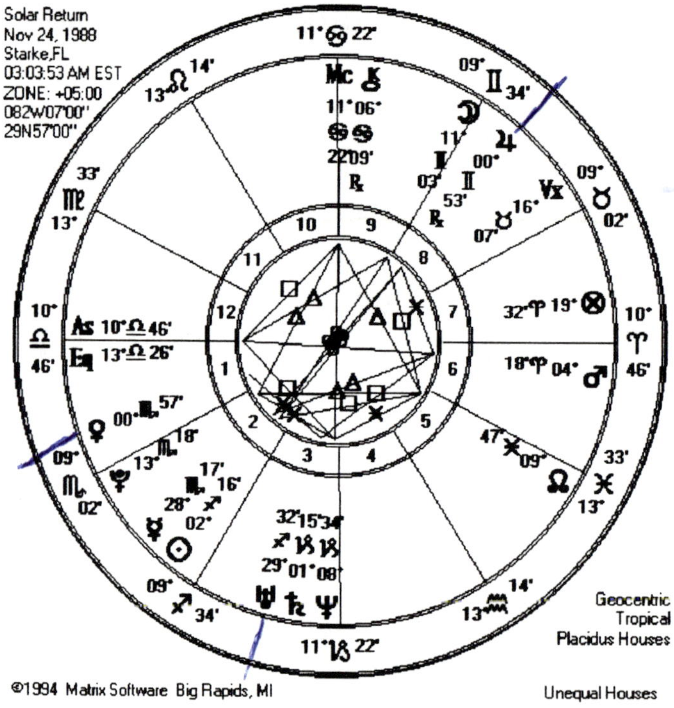

Chart 2 - Theodore Bundy's Solar Return Chart for 1988

The Yods or Fingers of Fate

A yod is found in Bundy's natal/progressed and solar return charts for 1988. In his natal/progressed, we find a yod consisting of a progressed Sun at 14:59 Capricorn in the fifth house, a natal North Node at 12:06 Gemini R in the tenth house, and a natal Pluto at 13:21 Leo R in the twelfth house. The yod in Bundy's solar return chart is found between his solar Venus at 00:57 Scorpio in the first house, a solar Uranus at 29:32 Sagittarius in the third house, and a solar Jupiter at 00:53 Gemini R in the eighth house.

Birth Data for Theodore Bundy

Birth date:	November 24, 1946
Birth time:	10:35 PM/EST
Birth place:	Burlington, VT
Longitude:	073W13 Latitude: 44N29
Source:	*Astro-Data III* by Lois R. Rodden, page 34

Crime Charts

When examining Crime Charts, astrologers are sometimes faced with the preliminary problem of determining whether a crime has actually been committed. There are other deaths which have been ruled accidental and may still appear to be intentional. One example of this is the death of Karen Silkwood. For more than twenty-five years, it has been argued both ways, and her death is still shrouded in mystery. By examining her astrological chart for that fateful day in 1974, it is possible to find additional clues that can help solve this mystery.

Picture 1 - Picture from Silkwood's security badge to enter Kerr-McGee

Karen Gay Silkwood
A Criminal Case?

Karen Silkwood's family members are still haunted by the circumstances surrounding her death. Silkwood was 26, and a divorced mother with three young children, when she joined Kerr McGee Corporation on August 5, 1972. She was excited at the prospect of working as a laboratory analyst at the nuclear-fuel processing plant near Crescent, OK. She was going to earn $3.45 an hour grinding and polishing plutonium.

Silkwood had worked at Kerr-McGee for about two years when she became involved in its union. After complaining to the union about a broken oxygen tank that was brought to revive a coworker who fainted, the union immediately recruited her to represent them in their upcoming contract talks. Coinciding with her union involvement, Silkwood became contaminated with plutonium during two accidents. One involved an improperly cleaned vacuum and the other involved a leaking glove box.

The union then sought Silkwood's help in collecting information on other safety violations and problems in the Kerr-McGee plant. There were 39 specific items on her list when she and other activists from her local met with representatives of the Oil, Chemical, and Atomic Workers Union in Washington, DC. Union officials needed to secure evidence of these violations and enlisted Silkwood to obtain the radiographs of "doctored" fuel pellets and photocopies of incriminating documents. She was to turn this evidence over to a union representative in about two months, and he agreed to make it available to Steve Wodka, a *New York Times* reporter.

Shortly before Silkwood was to meet with the union representative, she again became involved in several contamination incidents. The sources for these incidents could not be readily determined. However, when the company decontamination squad went to her home they found high levels of plutonium in her bathroom, bedroom and kitchen--with the highest readings coming from the food in her refrigerator. Silkwood was monitored for internal-body contamination at the Los Alamos Scientific Laboratory and returned to work a week later.

After gathering the requested evidence, Silkwood met with her local union on November 13, 1974. She left this meeting carrying a manila folder containing the evidence which she planned to give to the New York Times reporter. On her way to that meeting, Silkwood's car careened off Route 74 outside of Oklahoma City. She was killed instantly. The manila folder containing the evidence was not in her car nor was it ever found.

Investigations into Silkwood's death determined it to be accidental. However, many thought that someone wanted her silenced. A private auto-accident investigator hired by Silkwood's family theorized that her car may have been forced off the road by another vehicle.

In August 1986, after a ten-year plutonium contamination lawsuit, Kerr-McGee Corporation agreed in an out-of-court settlement to give the estate of Karen Silkwood $1.38 million. However, Kerr-McGee in no way accepted any blame for the alleged plutonium contamination of Silkwood.

In 1994, Silkwood's family attempted to retrieve her tissue samples and several bone fragments left behind from post-mortem testing at the Los Alamos National Laboratory in New Mexico. Although the laboratory has stated that it had permission to remove Silkwood's bones and some of the organs, including the brain, her parents said they never gave their permission. The family said that they did not know these body parts had been removed and taken to Los Alamos until during the trial against Kerr-McGee in 1979. The lab has since contacted the Silkwoods to find out what they wished to have done with their daughter's remains.

/WORLD Essex County Newspapers, Saturday, August 23, 1986 A3

Company settles in Silkwood suit

OKLAHOMA CITY (AP) — Kerr-McGee Corp. agreed Friday to give the estate of Karen Silkwood $1.38 million to settle a 10-year-old plutonium contamination lawsuit against the energy company.

Ms. Silkwood, a lab analyst at the Kerr-McGee Cimarron Plutonium Plant near Crescent, died in an auto accident Nov. 13, 1974. The 28-year-old employee was on her way to meet with a New York Times reporter to discuss alleged safety problems at the plant.

A federal court jury in 1979 awarded her family $10.5 million in punitive and compensatory damages, but the 10th U.S. Circuit Court of Appeals in Denver twice reversed the $10 million in punitive damages and last year ordered a new trial. That trial was to be limited to conduct by Kerr-McGee that might have allowed plutonium to escape from its Crescent plant and punitive damages that might result from that conduct.

With the settlement, Kerr-McGee in no way accepted blame for alleged plutonium contamination of Ms. Silkwood and in fact "believe we would have prevailed" in a new trial, company spokesman Rick Pereles said.

But Pereles said Kerr-McGee felt it was to its economic benefit to settle out of court and get the case behind it.

"That is not our only consideration, but it was a large consideration," Pereles said. He said the company believed "the economic decision was foremost for us ... The legal process would have cost in excess of the settlement."

The family's attorney, Jim Ikard, said Ms. Silkwood's three children on Thursday waived any future liability against the energy company.

Ikard said the three children will each receive about one-third of approximately $500,000. Most of the rest will go for attorney fees and expenses, including about $70,000 to Ms. Silkwood's father, Bill, as administrator of the estate.

Pereles said a settlement was agreed to after several months of negotiations between Kerr-McGee and attorneys for the family.

"Today marks an end in the chapter in the history of Kerr-McGee that has been difficult," he said. "By putting this matter behind us, we now focus our effort to improve the business results of Kerr-McGee."

Illustration 3 - Essex County Newspapers, August 23, 1986 (Silkwood)

Karen Silkwood's Natal Chart

In Karen Silkwood's natal chart, we find Jupiter at 27:15 Libra R in the first house of self inconjunct the Vertex at the ominous degree of 29:20 Taurus in the eighth house other's resources. She also has a natal yod comprised of Mercury at 08:21 Pisces in the fifth house of children and romantic relationships, Pluto at 10:11 Leo R in the tenth house of career and authority figures and Neptune at 08:10 Libra R in the twelfth house of secret enemies and self-undoing.

The Progressed Chart for 1974

In Silkwood's progressed chart, the progressed Moon at 25:51 Libra is in the natal first house of self and is approaching the critical degree of 26 Libra. Progressed Pluto at the critical degree of 09:40 Leo R in the natal tenth house of career and persons of authority. The progressed Moon and Pluto form a yod with natal Mercury at 08:21 Pisces in the fifth house of children and creativity and progressed Neptune at 07:29 Libra R in the twelfth house of secret enemies.

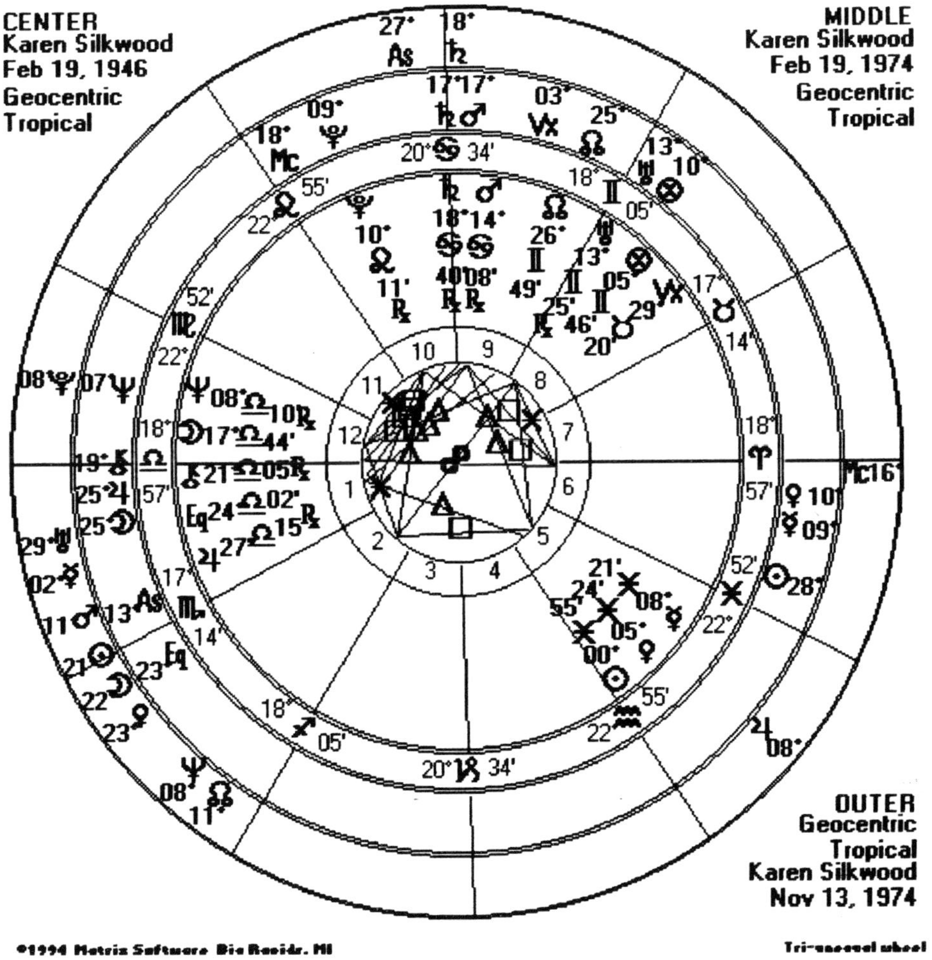

Chart 3 - Karen Silkwood's Three-ring Chart with Natal, Progressed Chart for 1974 and Transits for November 13, 1974

The Transiting Planets for November 13, 1974

Examining Silkwood's transits for the day of her death, we find transiting Uranus at 29:34 Libra in the natal first house of self. Transiting Uranus forms a yod with progressed Sun at 28:56 Pisces in the natal sixth house of work and service and the natal Vertex at 29:20 Taurus in the natal eighth house of other's resources. The transiting Part of Fortune is approaching the 29th degree at 28:59 Cancer in the natal tenth house of authority figures and public recognition.

A Total Solar Eclipse on June 20, 1974 occurred at 28:30 Gemini. The eclipse falls in Silkwood's ninth house of abstract mind and can indicate unforeseen circumstances that interrupt or cancel travel and education plans.

The Solar Return Chart for 1974

Silkwood's solar return chart for 1974 shows a yod formed by Uranus at 27:37 Libra R in the solar third house of everyday environment and travels. It sextiles the North

Node at 25:16 Sagittarius R in the solar fifth house and inconjuncts Mars at 25:55 Taurus in the solar tenth house.

The Fixed Star Algol is at 25:03 Taurus and conjuncts solar Mars at 25:55 Taurus in the solar tenth house of career and authority figures. Algol has a Saturn, Mars, Uranus, and Pluto influence. Algol is said to be the most violent star, and it is connected with strangulation, beheading, or other violent injury to the head, throat and neck. The degree for Algol is the January 1, 1920 placement taken from *The Fixed Stars & Constellations in Astrology* by Vivian E. Robson.

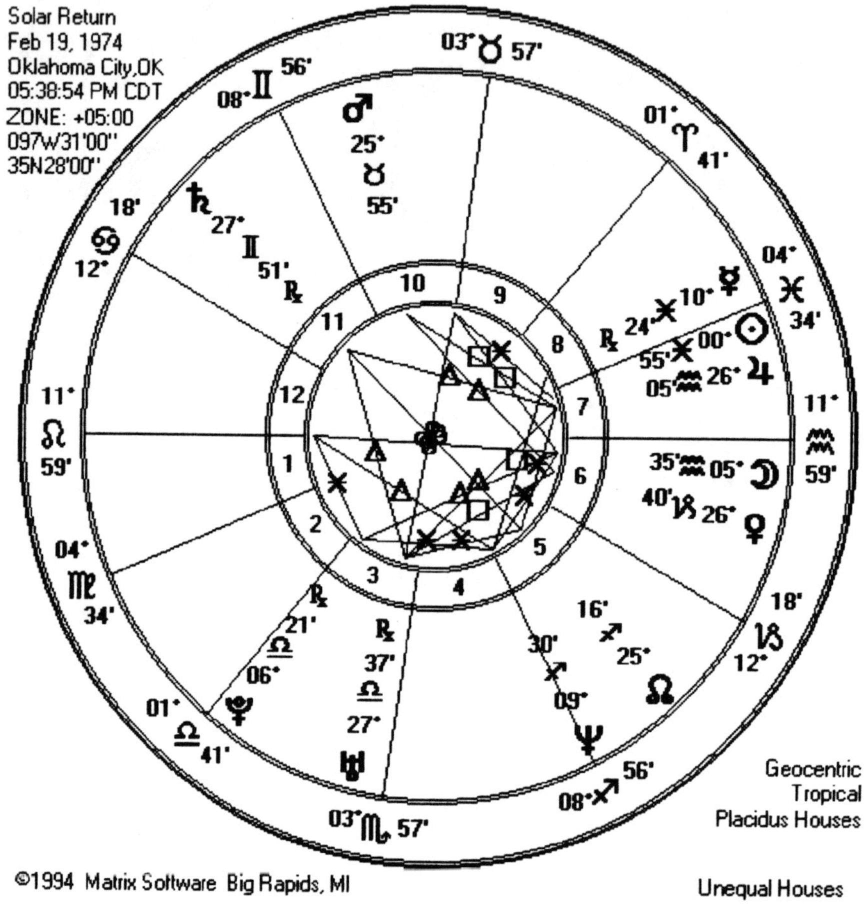

Chart 4 - Karen Silkwood's Solar Return Chart for 1974

The Yods or Fingers of Fate

In addition to Silkwood's natal and solar yods, another yod was being activated at the time of her death by natal Mercury at 08:21 Pisces and transiting Jupiter at 08:10 Pisces both in the fifth house of children, creativity and risk taking, progressed Pluto at 09:40 Leo in the tenth house of authority figures and progressed Neptune at 07:29 Libra R in the natal twelfth house of secret enemies. Triggering this yod were the transits of Neptune at 08:40 Sagittarius in the natal second house, Jupiter at 08:10 Pisces in the natal fifth house and Pluto at 08:18 Libra in the natal twelfth house.

We also find Silkwood's natal Vertex at 29:20 Taurus in the natal eighth house and ruling conditions surrounding one's death. This ominous degree forms a yod with her transiting Uranus at 29:34 Libra in the natal first house of self and progressed Sun at 28:56 Pisces in the natal sixth house of service and work.

On that fateful day, recall that Silkwood's mission was to meet with a newspaper reporter regarding the contamination incidents including her own exposure to radioactivity at Kerr-McGee. She intended to reveal evidence of the missing plutonium in addition to the falsified safety records. The yod in her charts with their 29 degree placements on that day offer clues to Silkwood's destiny.

The conditions surrounding Silkwood's death are ruled by the natal Vertex at 29:20 Taurus in the natal eighth house and involve transiting Uranus at 29:34 Libra in the natal first house ruling one's actions relating to a personal attitude, a sense of justice and an unexpected accident. The progressed Sun at 28:56 Pisces represents one's life and also health, chemical and chemical laboratories, secret matters, conspiracies and secret enemies. In Silkwood's case, her secret enemies were brought about by the natal sixth house which ruled her working conditions.

Basic House Interpretations

First House	Self	Seventh House	Others
Second House	Resources and values	Eighth House	Other's resources
Third House	Everyday environment	Ninth House	Abstract mind
Fourth House	Home	Tenth House	Career
Fifth House	Children	Eleventh House	Friendships
Sixth House	Work and service	Twelfth House	Secret enemies

A basic interpretation of the houses being activated by the yods in Silkwood's charts indicate that her employers and her working conditions would affect both her life and that of her children. Her values were attributable to her becoming interested in the working conditions and her involvement with the union at Kerr-McGee. This involvement could cause her to suffer through the actions of her secret enemies.

Astro*Carto*Graphy

A horoscope is a map of the sky at the moment of one's birth. Astro* Carto*Graphy takes astrology a step further because it transcends the relationship of the person to the nation. Astro*Carto*Graphy uses a map to indicate which parts of the world, cities and directions are in harmony or disharmony with a person's life. Cyclo*Carto*Graphy represents the modifications to the map over time through progressions and transits.

Silkwood's progressed Cyclo*Carto*Graphy for 1974 has aspects from progressed Saturn running through Oklahoma City in conjunction with progressed Mars. Saturn lines denote challenges and power struggles. The Mars lines indicate an area where there is a danger from attacks and accidents.

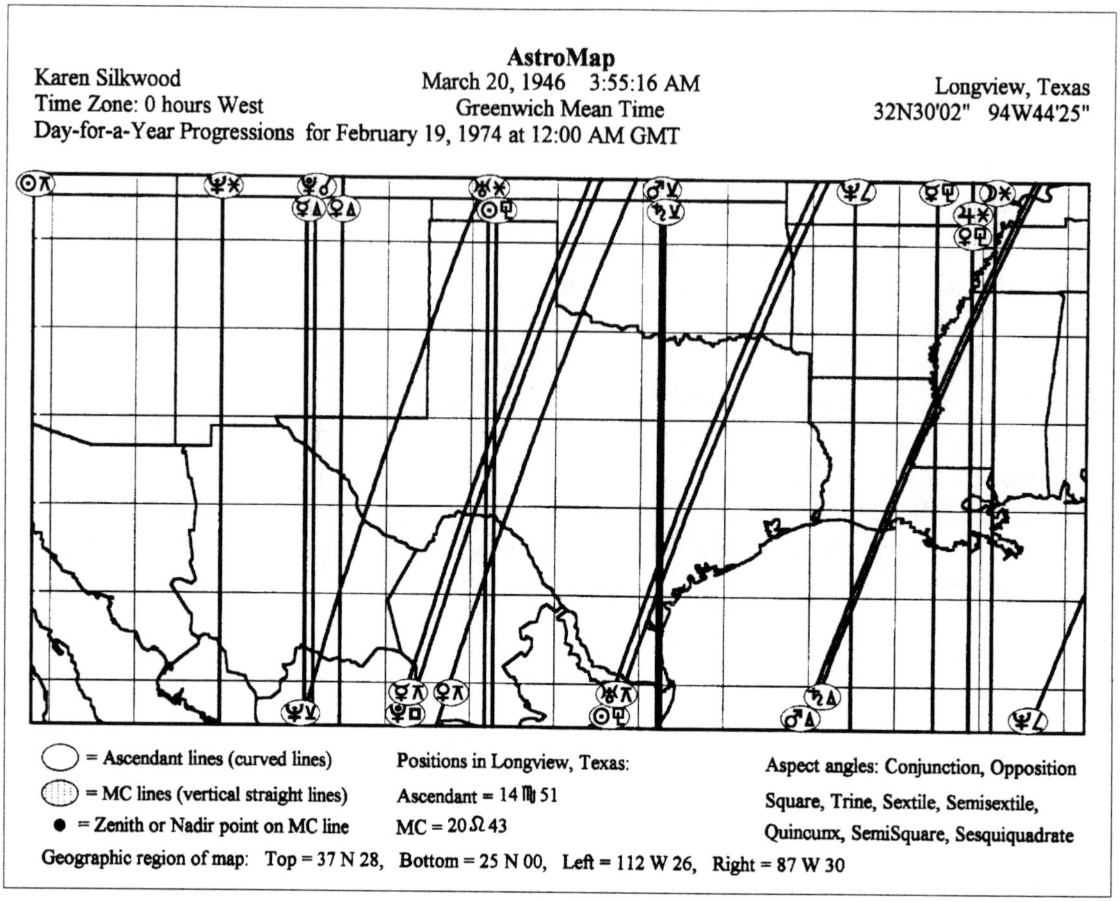

Map 1 - Karen Silkwood's Progressed AstroMap for 1974

In Conclusion

Even today, many questions regarding Silkwood's plutonium contamination and her death remain unanswered. Bill Silkwood, Silkwood's father, is convinced that someone killed his daughter. His standing $10,000 reward for information leading to an arrest and conviction in his daughter's death has gone unclaimed. However, he still has hopes that circumstances will change and someone will come forward.

In a personal telephone conversation with the Silkwoods in November 1997, Bill Silkwood mentioned an interesting fact in this case. The plutonium found in Silkwood's home was traced to a lot that she had never worked with at the Kerr-McGee plant. This means that someone else had brought the plutonium into her home.

Birth Data for Karen Silkwood

Birth date:	February 19, 1946		
Birth time:	9:50 PM CST		
Birth place:	Longview, TX		
Longitude:	094W44	Latitude:	32N30
Source:	*Astro-Data V* by Lois M. Rodden, page 139		

Placements and degrees of Karen Silkwood's charts:
Natal Placements

Moon	17:44 Lib	12th	Neptune	08:10 Lib R	12th
Sun	00:55 Pis	05th	Pluto	10:11 Leo R	10th
Mercury	08:21 Pis	05th	North Node	26:49 Gem R	09th
Venus	05:24 Pis	05th	Vertex	29:20 Tau	08th
Mars	14:08 Can R	09th	East Point	24:02 Lib	01st
Jupiter	27:15 Lib R	01st	Fortuna	05:46 Gem	08th
Saturn	18:40 Can R	09th	Midheaven	20:34 Can	
Uranus	13:25 Gem R	08th	Ascendant	18:57 Lib	

Eleventh house position is 22:55 Leo
Twelfth house position is 22:52 Vir

Second house position is 17:14 Sco
Third house position is 18:05 Sag

Critical Degrees: None
To be noted: Sun at the 00 degree and Vertex at the 29th degree
Intercepted Signs: None

Progressed Placements for 1974

Moon	25:51 Lib	01st	Neptune	07:29 Lib R	12th
Sun	28:56 Pis	06th	Pluto	09:40 Leo R	10th
Mercury	09:28 Ari R	06th	North Node	25:20 Gem R	09th
Venus	10:20 Ari	06th	Vertex	03:30 Can	09th
Mars	17:51 Can	09th	East Point	23:25 Sco	02nd
Jupiter	25:20 Lib R	01st	Fortuna	10:01 Gem	08th
Saturn	17:57 Can R	09th	Midheaven	18:36 Leo	10th
Uranus	13:45 Gem	08th	Ascendant	13:07 Sco	01st

Critical Degrees: Moon approaching 26:00 Libra in the first house of self, Pluto in the tenth house ruling authority figures
To be noted: Sun approaching the 29th degree of Pisces, ruler of the twelfth house of secret enemies, in the sixth house of work

Transiting Placements for November 13, 1974

Moon	22:54 Sco	02nd	Uranus	29:35 Lib	01st
Sun	21:23 Sco	02nd	Neptune	08:41 Sag	02nd
Mercury	02:59 Sco	01st	Pluto	08:18 Lib	12th
Venus	23:18 Sco	02nd	North Node	11:07 Sag R	02nd
Mars	11:27 Sco	01st	Vertex	08:37 Sco	02nd
Jupiter	08:11 Pis	05th	East Point	14:24 Can	09th
Saturn	18:44 Can R	09th	Fortuna	28:59 Can	10th

Critical Degrees: Sun
To be noted: Uranus at the 29th degree, Fortuna / Misfortuna approaching 29 degrees Cancer

Solar Return Placements for 1974 (Oklahoma City, OK)

Moon	05:35 Aqu	06th	Uranus	27:37 Lib R	03rd
Sun	00:55 Pis	07th	Neptune	09:30 Sag	05th
Mercury	10:24 Pis R	08th	Pluto	06:21 Lib R	03rd
Venus	26:40 Cap	06th	North Node	25:16 Sag R	05th
Mars	25:55 Tau	10th	Vertex	24:52 Sag	05th
Jupiter	26:05 Aqu	07th	Midheaven	03:57 Tau	
Saturn	27:51 Gem R	11th	Ascendant	11:59 Leo	

Eleventh house position is 08:56 Gem
Twelfth house position is 12:18 Can
Second house position is 04:34 Vir
Third house position is 01:41 Lib

Critical Degrees: Venus
To be noted: Sun at 00 degree
Intercepted Signs: None

Doreen Marciel
A Senseless Stabbing

In March 1992, Doreen Marciel experienced what most women consider to be one of life's worst nightmares. Marciel, a petite, attractive brunette of 45, agreed to help out a friend by working a couple of nights a week in the Kwik Mart on U.S. Route 1 in Fort Pierce, FL.

Picture 2 - Doreen Marciel

Marciel was working alone at about 8:00 p.m. on March 19, 1992 when a man in his twenties came into the mart for the third time that night. The first time, the young black man had shown her a parole card as proof of age to buy beer. After the second time, she mentioned to another customer who she knew that she thought the young man might be on drugs.

The last time, the man came into the store, he came up to the register and said that he wanted the cash. He then jumped over the counter and produced a large knife. He threatened Marciel by saying that either she give him the money or he would take her into the back room and rape her. She hesitated and as he grabbed for her, she backed away and began screaming. Marciel was no match for his strength as he attacked her. She continued to struggle against him while he viciously stabbed and slashed her.

That night, Marciel spent over twelve hours in the hospital's operating room as a result of what police called one of their city's most brutal attacks. The hospital drawing shows the slashes across her face. There were 16 wounds that required nearly 400 stitches to close them. Her left eye suffered extensive damage for which she has needed to undergo reconstructive plastic surgery. There were also numerous defensive wounds to her arms and hands.

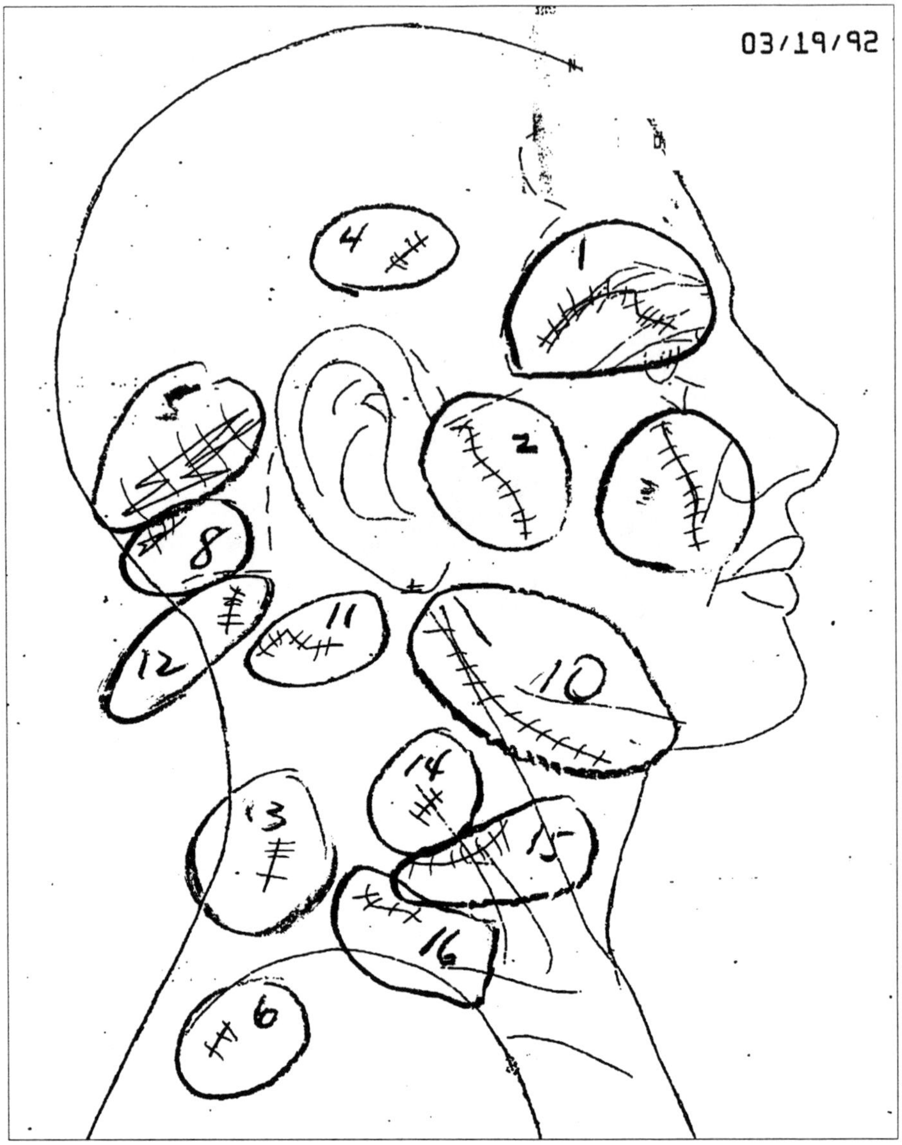

Illustration 4 - Hospital Drawing of Stab Wounds (Marciel)

Although Marciel provided a description of her attacker, the police search was hindered by the fact that the Kwik Mart did not have a video camera. A video camera was required by city ordinance if two clerks were not on duty. However, the Mart's owners said they had only purchased the Mart about six months earlier and were never told about that requirement.

Alvin Green, 28, surrendered to the police two days later. He was charged with attempted murder and armed robbery. This was not the first time Green had been in trouble with the law. In 1990, he was found guilty of robbing a gas station and sentenced to three years on probation. His probation was revoked in 1991 when he was convicted of breaking into someone's home and stealing a TV and a microwave. This time he was sentenced to three and a half years of which he served about five months in prison and was released on January 14, 1992.

Man charged in stabbing has a record

Cheryl Smith
TRIBUNE STAFF WRITER

A man accused in the stabbing and robbery of a Port St. Lucie woman last week had been released from prison in January after serving about 12 percent of his sentence.

Alvin L. Green, 28, of Revels Lane, Fort Pierce, served five months of a 3½-year prison term on two different criminal cases that involved charges of grand theft, burglary of a dwelling and violating his probation on a previous robbery conviction.

Green now faces attempted murder and armed robbery charges in the slashing and robbery

Green allegedly cut across the face and chest about 16 times, requiring nearly 400 stitches to close, her family said. She suffered a serious wound to the eye and was treated in the emergency room at HCA Lawnwood Regional Medical Center for nearly 12 hours.

In September 1990, he was convicted of a robbery at the King Orange Amoco, 7301 Orange Ave. He was living in Stuart and working as a landscaper at the time.

Green was found guilty and sentenced to three years on probation. But that probation was revoked in August 1991, when he was living in Port St. Lucie and was convicted of breaking into his Southwest Idaho Lane neighbors' home and stealing the womens' TV and microwave oven. Green was then working as a plasterer.

Having violated his probation and committed a new crime, Green was sentenced to 3½ years. But the sentences were to run concurrently, not consecutively. He served about five months in prison — with credit for the 97 days he spent in the county jail — before being released on Jan. 14.

Illustration 5 - Fort Pierce Tribune (Marciel)

Doreen Marciel's Natal Chart

In Marciel's natal chart, we find several sextiles and trines. Natal Mars at 22:50 Capricorn in the third house rules everyday environment. It is a few minutes away from an exact sextile with Jupiter at 22:53 Scorpio in the first house of self giving her vitality. Uranus at 18:25 R Gemini in the eighth house of other's possessions closely sextiles the Midheaven at 18:34 Leo, ruler of her house of career and public exposure.

We find two yods in Marciel's natal chart. The first comprises the planets of the Moon at 19:25 Scorpio in the first house of self, Mercury at 20:46 Capricorn in the third house of everyday environment, and Uranus at 18:25 R Gemini in the eighth house of other's possessions. The second yod is formed by Mercury at 20:46 Capricorn in the third house of everyday environment, Uranus at 18:25 R Gemini in the eighth house of other's possessions, and the Midheaven at 18:34 Leo. It should also be noted that the Fortuna or Misfortuna is placed at the critical degree of 04:18 Virgo in the tenth house ruler of her career and public recognition.

The Annular Eclipse occurred on January 04, 1992 at the critical degree of 13:51 Capricorn in the third house of everyday environment. This eclipse forms an inconjunct with natal Pluto at 12:35 R Leo in the ninth house.

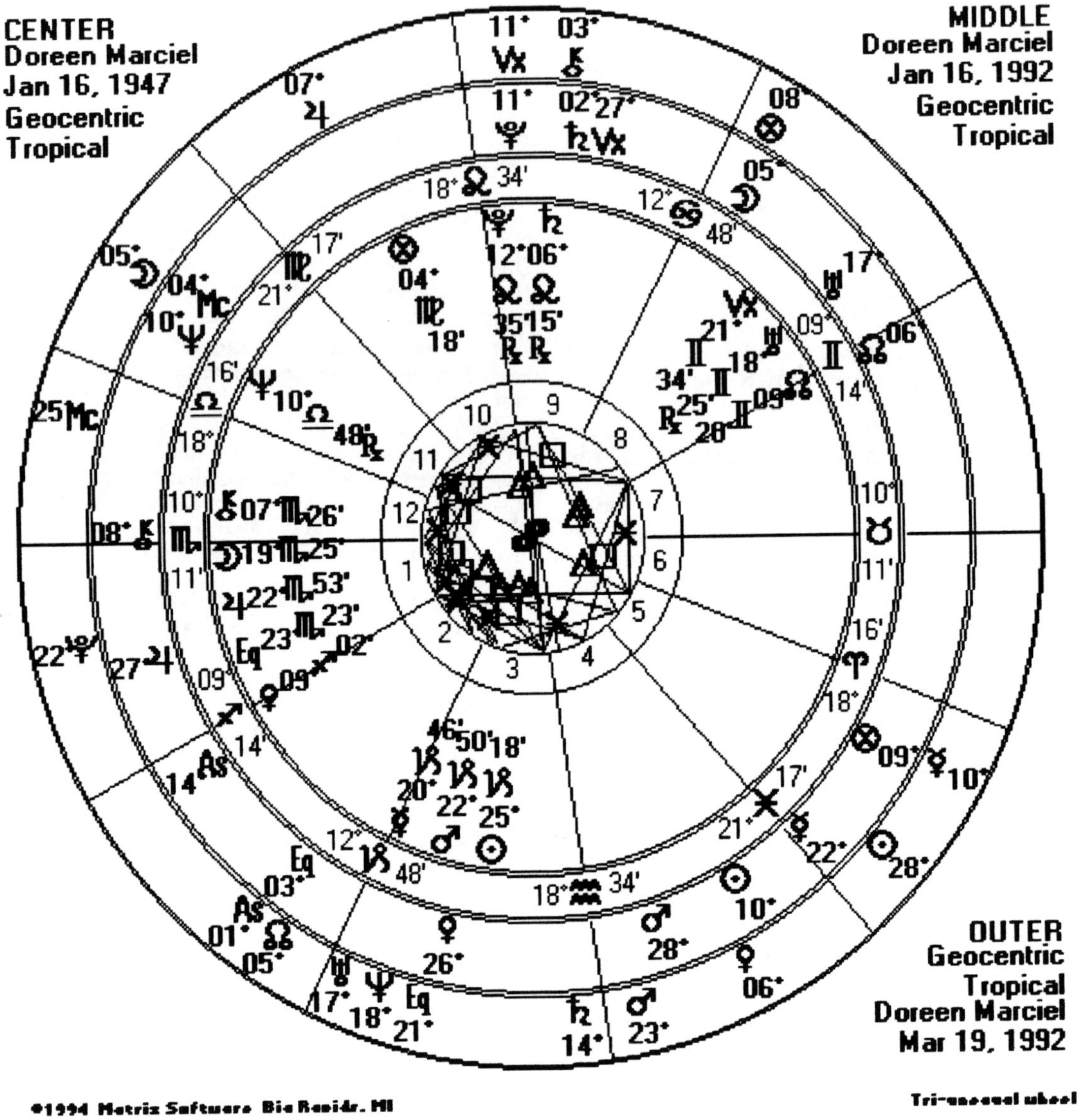

Chart 5 - Doreen Marciel's Three-ring Chart with Natal, Progressed Chart for 1992 and Transits for March 19, 1992

The Progressed Chart for 1992

In Marciel's progressed chart Venus at the critical degree of 26:38 Capricorn in the natal third house of everyday environment and Uranus at the critical degree of 17:45 Gemini in the natal eighth house. We also note that the progressed Sun at 10:51 Pisces in the natal fourth house forms a trine to the natal Ascendant at 10:11 Scorpio, ruler of knives and stabbing. There is also a tight yod which includes this Scorpion Ascendant. The yod encompasses the natal Ascendant at 10:11 Scorpio, progressed Fortuna or Misfortuna at 09:08 Aries, ruler of the head in the natal fifth house, and the natal North Node at 09:20 Gemini R in the natal eighth house. Gemini is the ruler of young men.

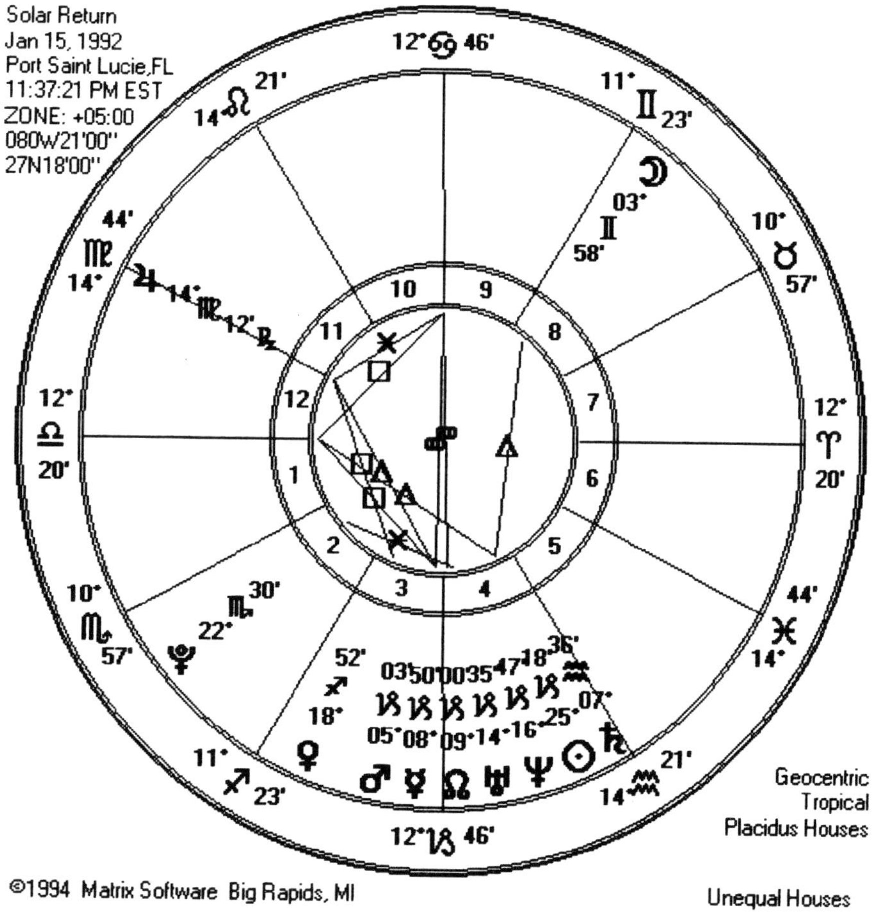

Chart 6 - Doreen Marciel's Solar Return Chart for 1992

The Solar Return Chart for 1992 (Port St. Lucie, FL)

Marciel's solar return chart for 1992 has Part of Fortune at the critical degree of 21:00 Aquarius in the fifth house. The solar Midheaven at 12:46 Cancer is fourteen minutes away from the critical degree of 13:00 Cancer. The solar Midheaven receives several oppositions from Mercury and North Node in the third house and Uranus and Neptune in the fourth house. The Ascendant at 12:20 Libra nears the critical degree of 13 Libra and receives squares from the same third and fourth house solar planets. The solar Jupiter at 14:12 Virgo R in the eleventh house has retrograded over the cusp from the twelfth house of hospitalization. Jupiter, the protector, is aspected by trines from the same planets that form the mentioned squares and oppositions.

In Conclusion

Since her attack, Marciel has had repeated reconstructive surgeries on her face and eye. Her attacker, Alvin Green, is again serving his time in jail. In an effort to prevent people like Green from committing other crimes, Stop Turning Out Prisoners (STOP), a citizens group fighting the early release program, featured him on a poster that was hung throughout the neighboring counties.

Birth Data for Doreen Marciel:

Birth date:	January 16, 1947		
Birth time:	01:40 AM/EST		
Birth place:	Queens, NY		
Longitude:	073W52	Latitude:	40N43
Source:	Birth Certificate		

Placements and degrees of Doreen Marciel's charts:
Natal Placements

Moon	19:25 Sco	01st	Neptune	10:48 Lib R	11th
Sun	25:18 Cap	03rd	Pluto	12:35 Leo R	09th
Mercury	20:46 Cap	03rd	North Node	09:20 Gem R	08th
Venus	09:02 Sag	01st	Vertex	21:34 Gem	08th
Mars	22:50 Cap	03rd	East Point	23:23 Sco	01st
Jupiter	22:53 Sco	01st	Fortuna	04:18 Vir	10th
Saturn	06:15 Leo R	09th	Midheaven	18:34 Leo	
Uranus	18:25 Gem R	08th	Ascendant	10:11 Sco	

Eleventh house position is 21:17 Vir
Twelfth house position is 18:16 Lib
Second house position is 09:14 Sag
Third house position is 12:48 Cap

Critical Degrees: Fortuna or Misfortuna in the tenth house of employer

To be noted: None
Intercepted Signs: None

Progressed Placements for 1992

Moon	05:46 Can	08th	Neptune	10:11 Lib R	11th
Sun	10:51 Pis	04th	Pluto	11:33 Leo R	09th
Mercury	22:38 Pis R	05th	North Node	06:57 Gem R	07th
Venus	26:38 Cap	03rd	Vertex	27:39 Can	09th
Mars	28:06 Aqu	04th	East Point	03:28 Cap	02nd
Jupiter	27:20 Sco	01st	Fortuna	09:08 Ari	05th
Saturn	02:53 Leo R	09th	Midheaven	04:07 Lib	11th
Uranus	17:45 Gem	08th	Ascendant	14:13 Sag	02nd

Critical Degrees: Venus is in the third house of everyday environment and Uranus is in the eighth house of shared resources
To be noted: None

Transiting Placements for March 19, 1992

Moon	05:55 Lib	11th	Uranus	17:32 Cap	03rd
Sun	28:55 Pis	05th	Neptune	18:40 Cap	03rd
Mercury	10:58 Ari R	05th	Pluto	22:48 Sco R	01st
Venus	06:31 Pis	04th	North Node	05:39 Cap R	02nd
Mars	23:11 Aqu	04th	Vertex	11:52 Leo	09th
Jupiter	07:16 Vir R	10th	East Point	21:28 Cap	03rd
Saturn	14:46 Aqu	03rd	Fortuna	08:12 Can	08th

Critical Degrees: None
To be noted: The Sun approaching the 29th degree

Solar Return Placements for 1992 (Port St. Lucie, FL)

Moon	03:58 Gem	08th	Neptune	16:47 Cap	04th
Sun	25:18 Cap	04th	Pluto	22:30 Sco	02nd
Mercury	08:50 Cap	03rd	North Node	09:00 Cap R	03rd
Venus	18:52 Sag	03rd	Vertex	03:25 Gem	08th
Mars	05:03 Cap	03rd	East Point	15:04 Lib	01st
Jupiter	14:12 Vir R	11th	Fortuna	21:00 Aqu	05th
Saturn	07:36 Aqu	04th	Midheaven	12:46 Can	
Uranus	14:35 Cap	04th	Ascendant	12:20 Lib	

Eleventh house position is 14:21 Leo
Twelfth house position is 14:44Vir
Second house position is 10:57 Sco
Third house position is 11:23 Sag

Critical Degrees: Fortuna and the Midheaven approaching 13:00 Can
To be noted: None
Intercepted Signs: None

MISSING PERSONS

Karen Wilson
Without A Clue

Karen Wilson was a senior and an honor student at the State University of New York at Albany when she disappeared on March 27, 1985. She had been looking forward to spending her spring break in Florida with some of her classmates. Wilson who was 5'3" and weighed 115 pounds, had just turned 22 a month earlier. In addition to attending school, she had worked for about a year as a New York State Assembly intern to Assemblyman Samuel Colman. Before leaving work on the Wednesday that she disappeared, Wilson changed into blue jeans and a blue, short-sleeved blouse so that she could stop at the Tanning Hut in Colonie, NY to begin working on her tan.

Picture 3 - Karen Wilson

Wilson was not seen again after she left work. It is known, however, that she went to the Colonie Center Mall where bought some supplies for her office and T-shirts to take to Florida. She used Visa checks given to her by her parents and therefor the purchases were documented.

There were reports that Wilson left the Tanning Hut at about 7:30 p.m., however her mother has recently advised the writer that her daughter had an appointment at the

Tanning Hut, but that her name was never checked off as having shown up. All others who had appointments for that day had been checked off. In addition, reports by passersby that Wilson may have seen walking near the SUNY campus minutes after the time that she was to have left the Tanning Hut could never be confirmed with a positive identification.

Grim anniversary
Missing college student presents baffling case
Search for coed goes on
Police refuse to give up search for Karen W...
Wilsons on a tearful search
A year of devastating grief
Karen Wilson's parents still cling to the memories
THE TIMES UNION Albany, N.Y., Friday, March 27, 1992 B-3
...nps police after seven years

Karen Wilson's disappearance, said Wingate.

Detectives have also visited four prisons in the hunt for new informa-

Murray said that while the Wilson case has been a difficult one for police, experience shows that a cold trail can suddenly turn hot...

Continued from C-8

WILSON: Still no clues, 10 years later

been buoyed by the relentless efforts of UAlbany Police Chief Jim Williams and other police officers to solve the riddle of what happened to Karen Wilson.

When they were convinced that there was no hope of ever finding her alive, the Wilsons resolved to try to get on with their lives as best they could. "If you try to live it everyday for too long, it almost starts to destroy you," Jenny Wilson said in a telephone interview from her home in Redondo Beach, Calif., just south of Los Angeles. "We both try to think of her in happy terms. That's what

Raymond Wilson. "I would rather not know than see the person go unpunished."

Jenny Wilson said she and her husband have never even considered it a remote possibility that Karen would have run away.

Raymond Wilson remembered coming to Albany 10 years ago while hundreds of state troopers and local police officers searched the Six Mile Water Works and other areas near the campus. He said he asked one State Police supervisor about the chances of finding his daughter. "I told him to shoot straight with me,"

daughter of one of her friends, about the same age Karen would be now, just had her first child. "That kind of stuff impacts real hard," she said. "You hear that and you can't help but think about what she might be doing now."

Raymond Wilson added, "There's probably not an hour that goes by that you're not reminded of Karen by somebody saying something or something happening to remind you."

The Wilsons say they are not likely candidates to serve on juries, because criminal defense lawyers would immediately move to have them eliminated from any pool of prospective

liams has seen the heartbreak Wilsons up close. On a regular he confers with them by tele "It is getting increasingly frust because I don't have anythi substance to tell them," he said

Williams is optimistic that o there will be a breakthrough. day she will be found," he said I'm not going to give up unt occurs."

State Police Capt. Jeffrey who supervises criminal inv tions at Troop G headquart Loudonville, said detectives a tinually assigned to give the

Illustration 6 - Albany Times Union and Schenectady Gazette (Wilson)

Wilson's unused airline tickets sat on a desk in her dormitory room. The weeks went by, but Wilson's family, the investigators and her friends refused to give up in their search for her. An extensive search was conducted of the Six-Mile Waterworks and the surrounding area where she was last seen, but no clues were found. State police encouraged people to come forward with anything that might help them in their investigation and they followed up on the smallest of leads.

The Wilsons offered a $10,000 reward for information leading to the arrest and conviction of the person or persons responsible for their daughter's disappearance. Assemblyman Coleman, his staff and other members of the New York State Assembly formed the "Find Karen Wilson Fund" and SUNY students made door-to-door solicitations for funds in order to continue the search for her. Her boyfriend, collected additional funds to pay the expenses for the continuation of the five-week search, which had not produced any substantial leads. Missing Person posters were distributed across the country and national television networks were contacted to spark an interest in Wilson's disappearance. The Church of Latter Day Saints, where Wilsons were members mailed out flyers to each of its churches.

In March 1996, eleven years later, Wilson's disappearance still haunted James Williams, then Director of the State University at Albany Department of Public Safety. Over the years, he had personally followed up on every small lead. He also spent much of his spare time in the sun or the rain with a shovel in his hands, poking for anything that could be a clue, a piece of clothing, or a crude grave-site. For years, Williams called the Wilsons regularly to let them know that someone cared that their daughter is still missing.

Karen Wilson's Natal Chart

In Wilson's natal chart, we find Jupiter at the critical degree of 17:26 Pisces in the first house opposing Moon at 14:18 Virgo in the seventh house. Her Sun is at the critical degree of 21:10 Aquarius in the twelfth house. There is a T-square involving Mars at 12:32 R Leo in the sixth house of health and work, Saturn at 14:41 Aquarius in the twelfth house of institutions and confinement, and Neptune at 15:39 Scorpio in the eighth house.

The total Solar Eclipse at 00:50 Sagittarius of November 22, 1984 occurred in Wilson's ninth house. A solar eclipse in the ninth house can indicate unforeseen circumstances that interrupt or cancel travel and education plans.

The Progressed Chart

In Wilson's progressed chart for February 10, 1985, we find the progressed Moon at 07:20 Cancer in the natal fourth house, indicating changes in her personal affairs. There are no aspects from the progressed to progressed planets within one degree and there are no critical degrees.

However, several progressed planets aspect the natal planets. The progressed Ascendant at 20:57 Aries in the natal first house sextiles natal Sun at 21:10 Aries in the natal twelfth house. The progressed Midheaven at 11:09 Capricorn in the natal eleventh house trines natal Pluto at 11:22 Virgo in the natal seventh house. Lastly, the progressed Sun at 13:20 Pisces in the natal first house opposes natal Moon at 14:18 Virgo in the natal seventh house.

The Transiting Planets for March 27, 1985

Wilson had three challenging transiting aspects in an orb of one degree +/- on the evening that she disappeared. Transiting Uranus at the critical degree of 17:59 Sagittarius in the natal ninth house of higher mind separates from a square to natal Jupiter at the critical degree of 17:26 Pisces in the natal first house of self. Transiting Uranus also applies to a conjunction with the natal Midheaven at 18:59 Sagittarius, and transiting

Jupiter at 10:11 Aquarius in the twelfth house of self-undoing applies to an inconjunct with natal Pluto at 11:22 Virgo in the natal seventh house of one-to-one relationships.

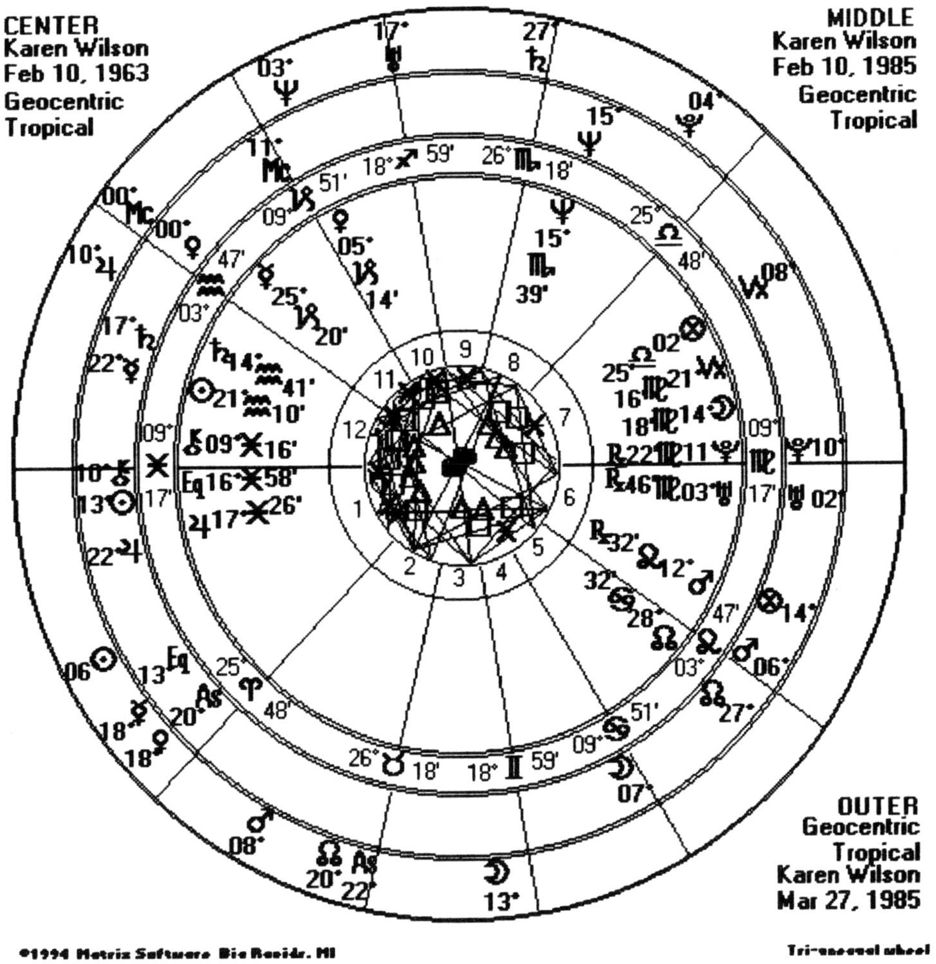

Chart 7 - Karen Wilson's Three-ring Chart with Natal, Progressed Chart for 1985 and Transits for March 27, 1985

Wilson was always known to be steadfast and very dependable, however Uranus square Jupiter suggests that she may have been overly optimistic or impulsive at this time. Unpredictable behavior is also indicated either on her part or on the part of another. This aspect occurs twice in the average lifetime. Uranus forming a conjunction with the Midheaven brings one unexpected changes and public recognition. Jupiter applying to an inconjunct with Pluto indicates that one can either perpetrate or be subject to power struggles. This can also suggest problems connected with sexual morality.

Wilson also has transiting Mars at 08:59 Taurus in the second house of personal values applying to a sextile with her natal Ascendant at 09:17 Pisces. The aggressiveness of Mars reaching the critical degree of 09:00 Taurus makes personal values a critical issue at that time.

A Total Solar Eclipse on November 22, 1984 occurred at 00:50 Sagittarius. The eclipse occurs in Wilson's ninth house and can indicate that travel and education plans may be interrupted due to unforeseen circumstances.

The Solar Return Chart for 1985 (Colonie, NY)

Wilson's solar return chart for 1985 has the 29th degree of Taurus on her Midheaven and the critical degree of 04:23 Virgo on her Ascendant. The solar Uranus at the critical degree of 17:15 Sagittarius in the fourth house ruling personal affairs. The solar Sun is at the critical degree of 21:10 Aquarius in the solar sixth house of work and personal service. Solar Jupiter is at the new start degree of 00:46 Aquarius in the fifth house of love relationships, forming a trine to the Midheaven or career area of her chart.

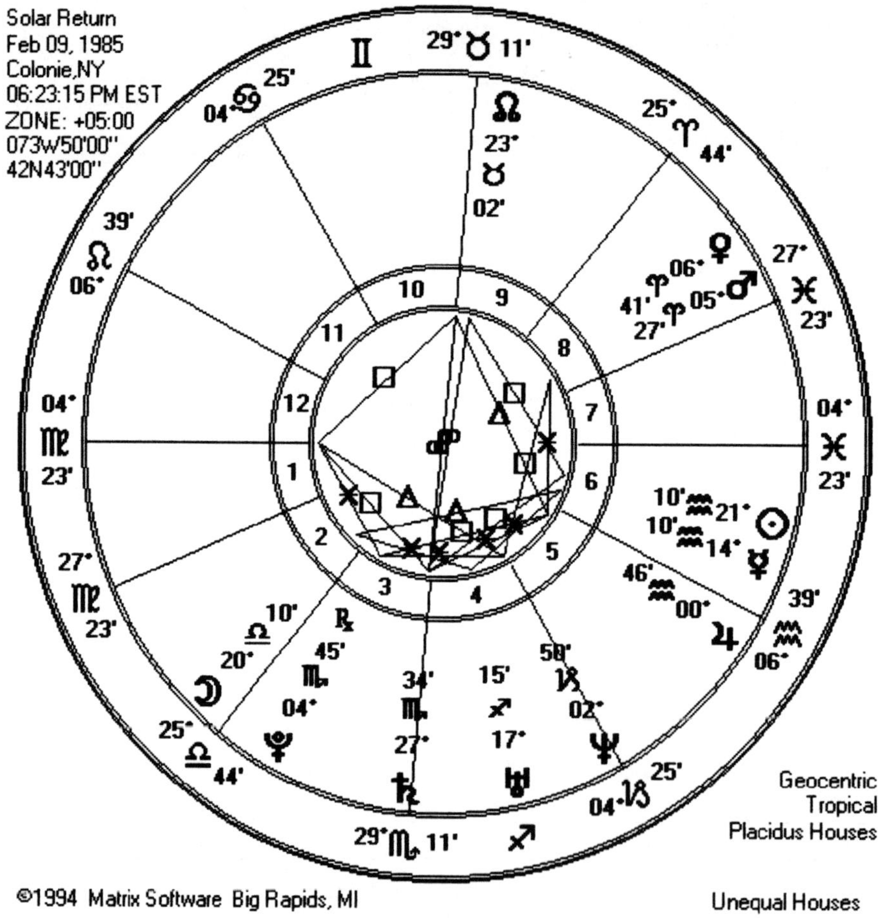

Chart 8 - Karen Wilson's Solar Return Chart for 1985

There are two closely formed yods in Wilson's solar return chart. They are formed by the Ascendant at 04:23 Virgo, Pluto at 04:45 Scorpio R in the third house of everyday environment and Mars at 05:27 Aries in the eighth house of shared values and resources. The second yod is formed by the Ascendant, Pluto and Venus at 06:41 Aries in the eighth house.

The Yods or Fingers of Fate

One of the yods in Wilson's charts is comprised of the progressed Ascendant at 20:57 Aries, transiting Mercury at 18:09 Aries, transiting Venus at 18:26 Aries in the natal first house of self, natal Vertex at 21:16 Virgo in the natal seventh house one-to-one relationships, and Natal Sun at 21:10 Aquarius in the critical degree and progressed Mercury at 22:51 both in natal twelfth house of possible self-undoing.

Another yod is comprised of the progressed East Point at the critical degree of 13:10 Aries in the natal first house of self, the natal Moon at 14:18 Virgo in the natal seventh house of one-to-one relationships, and natal Neptune at 15:39 Scorpio in the natal eighth house of shared resources.

Yod No. 1

First House	Progressed Ascendant	20:57 Aries
First House	Transiting Mercury	18:09 Aries
First House	Transiting Venus	18:26 Aries
Seventh House	Natal Vertex	21:16 Virgo
Twelfth House	Natal Sun	21:10 Aquarius - Critical Degree
Twelfth House	Progressed Mercury	22:51 Aquarius

The Yod includes the First, Seventh and Twelfth Houses

Yod No. 2

First House	Progressed East Point	13:10 Aries - Critical Degree
Seventh House	Natal Moon	14:18 Virgo
Eighth House	Natal Neptune	15:39 Scorpio
Eighth House	Progressed Neptune	15:35 Scorpio R

The Yod includes the First, Seventh and Eighth Houses

What follows is an interpretation of these two yods in Wilson's charts using repetitive Basic Keywords for each sign, house and planet. An excellent source of these keywords and what has been used for this example is *The Rulership Book* by Rex E. Bills.

Yod No. 1

First House, Progressed Mercury and Venus in Aries

First House	Progressed Mercury / Venus	Aries
Habits	Gasoline stations/workers	Instruments, cutting or sharp
One's walk	Pairs, things that come in pairs	Head, in general
	Automobiles, auto trips	
	Strangulation	
	Wednesday (Karen disappeared on Wed., March 27, 1985)	

Seventh House, Natal Vertex in Virgo

Seventh House
Bartenders, bars
Boy friends
Meetings between strangers
Quarrels
Sunset

Virgo
Farms, farming, farmers
Hay, places where stored
Dairies, dairy farmers
Restaurant /workers
Working classes

Twelfth House, Natal Sun (Critical Degree) and Mercury in Aquarius

Twelfth House
Crime, criminals
Drugs, in general
Males, in general
Kidnapping, kidnapers
Arguments
Murder
Strangulation
Secret love affairs

Natal Sun and Mercury
Bosses, executives
Government/State officials
Local Travel
Gasoline stations/workers
Vehicles, small and light
Pairs, things that come in pairs

Aquarius
Acquaintances
Automobile mechanics
dealers, workers
Friends
Legislature, legislatures or
legislative branches

Yod No. 2

First House, East Point in Aries

First House
Habits
One's walk

East Point

Aries
Instruments, cutting or sharp
Head, in general

Seventh House, Moon in Virgo

Seventh House
Arguments with enemies
Bartenders, bars
Boy friends
Common people
Meetings between strangers
One's mate
Outcome of opposition
Quarrels
Sunset

Moon
Aqueducts
Barmaids, bartenders
Cisterns
Places, damp, wet or boggy
Evening
One's grave
Working classes
Pipes, water
Places near water

Virgo
Crops, Crop growers
Farms, farming, farmers
Wheat fields
Hay, places where stored
Level lands, pastures
Dairies, dairy farmers
Restaurant /workers

Eighth House, Neptune in Scorpio

Eighth House
Death, manner of one's with 4th
Foreign persons
Friends in one's business or professional life
Life after death
Mortality Plight, one's
Reincarnation

Neptune
Alliances, covert
Asphyxiation
Aqueducts
Bartenders
Blackmail
Disappearances
Drains
Drowning
Gasoline stations, owners/workers
Infidelity
Pipes, water
Places damp, wet, or boggy
Plumbers, plumbing, plumbing lines

Scorpio
Bogs
Cesspools
Ponds, stagnant or ill-smelling
Pools of water, stagnant
Sewers, sewage disposal
Ruins near water
Floods, places subject to

In Conclusion

Karen is still missing and the investigation into her disappearance remains open.

Birth Data for Karen Wilson

Birth date:	February 10, 1963		
Birth time:	08:20 AM/MST		
Birth place:	Ogden, UT		
Longitude:	111W58	Latitude:	41N13
Source:	Birth certificate		

Placements and degrees of Karen Wilson's charts:
Natal Placements

Moon	14:18 Vir	07th	Neptune	15:39 Sco	08th
Sun	21:10 Aqu	12th	Pluto	11:22 Vir R	07th
Mercury	25:20 Cap	11th	North Node	28:32 Can R	05th
Venus	05:14 Cap	10th	Vertex	21:16 Vir	07th
Mars	12:32 Leo R	06th	East Point	16:58 Pis	01st
Jupiter	17:26 Pis	01st	Fortuna	02:25 Lib	07th
Saturn	14:41 Aqu	12th	Midheaven	18:59 Sag	
Uranus	03:46 Vir R	06th	Ascendant	09:17 Pis	

Eleventh house position is 09:51 Vir
Twelfth house position is 03:47 Aqu
Second house position is 25:48 Ari
Third house position is 26:18 Tau

Critical Degrees: Sun, Jupiter, and the East Point approaching 17 degrees of Pisces
To be noted: None

Intercepted Signs: None

Progressed Placements for 1985

Moon	07.20 Can	04th	Neptune	15:35 Sco R	08th
Sun	13:20 Pis	01st	Pluto	10:49 Vir R	07th
Mercury	22:51 Aqu	12th	North Node	27:22 Can R	05th
Venus	00:11 Aqu	11th	Vertex	08:50 Lib	07th
Mars	06:16 Leo R	06th	East Point	13:10 Ari	01st
Jupiter	22:38 Pis	01st	Fortuna	14:57 Leo	06th
Saturn	17:15 Aqu	12th	Midheaven	11:09 Cap	11th
Uranus	02:49 Vir R	06th	Ascendant	20:57 Ari	02nd

Critical Degrees: East Point
To be noted: Venus at 00 degrees

Transiting Placements for March 27, 1985

Moon	13:40 Gem	03rd	Uranus	17:59 Sag R	09th
Sun	06:54 Ari	01st	Neptune	03:36 Cap	10th
Mercury	18:09 Ari R	01st	Pluto	04:05 Sco R	08th
Venus	18:26 Ari R	01st	North Node	20:37 Tau R	02nd
Mars	08:59 Tau	02nd	Vertex	24:01 Lib	07th
Jupiter	10:11 Aqu	12th	East Point	05:19 Tau	02nd
Saturn	27:47 Sco R	09th	Fortuna	29:06 Can	05th

Critical Degrees: Mars and Uranus
To be noted: East Point at 29 degrees Cancer

Solar Return Placements for 1985 (Colonie, NY)

Moon	20.10 Lib	02nd	Neptune	02:50 Cap	04th
Sun	21:10 Aqu	06th	Pluto	04:45 Sco R	03rd
Mercury	14:10 Aqu	06th	North Node	23:02 Tau R	09th
Venus	06:41 Ari	08th	Vertex	01:49 Aqu	05th
Mars	05:27 Ari	08th	East Point	24:40 Leo	12th
Jupiter	00:46 Aqu	05th	Fortuna	03:22 Tau	09th
Saturn	27:34 Sco	03rd	Midheaven	29:11 Tau	
Uranus	17:15 Sag	04th	Ascendant	04:23 Vir	

Diane Leslie Reeves
The Disappearance of Diane

Diane Reeves was 23 years old when she disappeared on the morning of July 26, 1989 from her home in Ferndale, Washington. Reeves picked up her part-Siamese kitten and left the trailer she shared with her boyfriend. He later told police that he heard her 1983 Subaru start and watched as she drove away. He also stated that he has not seen her since that fateful morning.

Reeves attended the University of Southern Mississippi on a full scholarship and graduated with honors in May 1988. Wanting a break from college and work, she accepted a summer job at Yosemite National Park in California. It was at Yosemite that she met and began dating her boyfriend. On October 5, 1988, they moved to Ferndale, where they bought a small. trailer. Reeves had a special love of nature and they both liked Ferndale because they could hike and backpack in the surrounding areas.

Picture 4 - Diane Reeves

At the time of her disappearance, Reeves was working for a utility company as a flagger. Police know that she worked from 8:00 a.m. to 12:00 p.m. on July 25th, but they found the other circumstances prior to her disappearance somewhat confusing. Stories from family and friends suggested different possible problems. One said there may have been a problem with Reeves's pap smear that she needed to follow-up. However, another family member advised that her doctor's appointment for a follow-up examination had not been kept.

Illustration 7 - Diane Reeves's Missing Person Flyer (Reeves)

Reeves had also suffered in the past from a kidney problem and this may have reoccurred. The clinic refused to release her medical records, even though the family was able to secure a court order for them. This could be indicative of the fact that in the state of Washington, medical records and any corresponding records cannot be divulged to anyone without the patient giving their prior written approval.

How Reeves spent the hours between noon on July 25 and 6:00 a.m. of the day she disappeared is also questionable and inconclusive. Her boyfriend told police that she had taken a shower at about 9:00 p.m. and later went for a walk, ending up at a nearby farmhouse. When the people at the farmhouse saw Reeves early the next morning, they reportedly asked her if she had been in an accident She said she didn't know. Sometime between 5:00 and 5:30 a.m. on the 26th, she called her boyfriend from the farmhouse and he picked her up. He reported that he asked her if she wanted to go to the hospital, but she said "no." He took her home and said that she took a shower and laid down for a short time on the couch.

At about 6:00 a.m., Reeves picked up her kitten and the car keys, and left the trailer. About 40 minutes later, a nurse reported that Reeves had stopped at her house. The nurse said she was awakened by the sound of Reeves walking on the gravel in the driveway and called out to ask her, "Do you want something?" The nurse said she had never seen Reeves before. Reeves told her she had left home because her boyfriend was on drugs and had kicked her out. She then asked if she could rest there in her car for awhile. The nurse reported that she sat in the car for about twenty minutes before she left the yard at about 6:45 a.m. At 10:00 a.m. on July 26, 1989, her boyfriend telephoned Reeve's sister-in-law and told her, "Diane is missing." He later filed a missing persons report with the sheriff's office in Bellingham at 4:32 p.m.

In the months following Reeves disappearance, there were several reported "sightings" of her, but all were unconfirmed. Two hunters found her car near the end of a logging road, eleven miles off the main highway on September 1, 1989 near Green Mountain which is in the Cascade Mountains. Later, her backpack and water bottle were also found in a nearby area.

Reeves has never contacted her family. However, shortly after her disappearance they reported several phone calls in which the caller did not respond. Since 1990, these unresponsive calls have ended. Although they have not heard from her, Reeves' family has not given up in their search for her. Her parents have spent much of their time traveling from Florida to Washington in order to search the areas where possible sightings have been reported and in the surrounding area where her car was found.

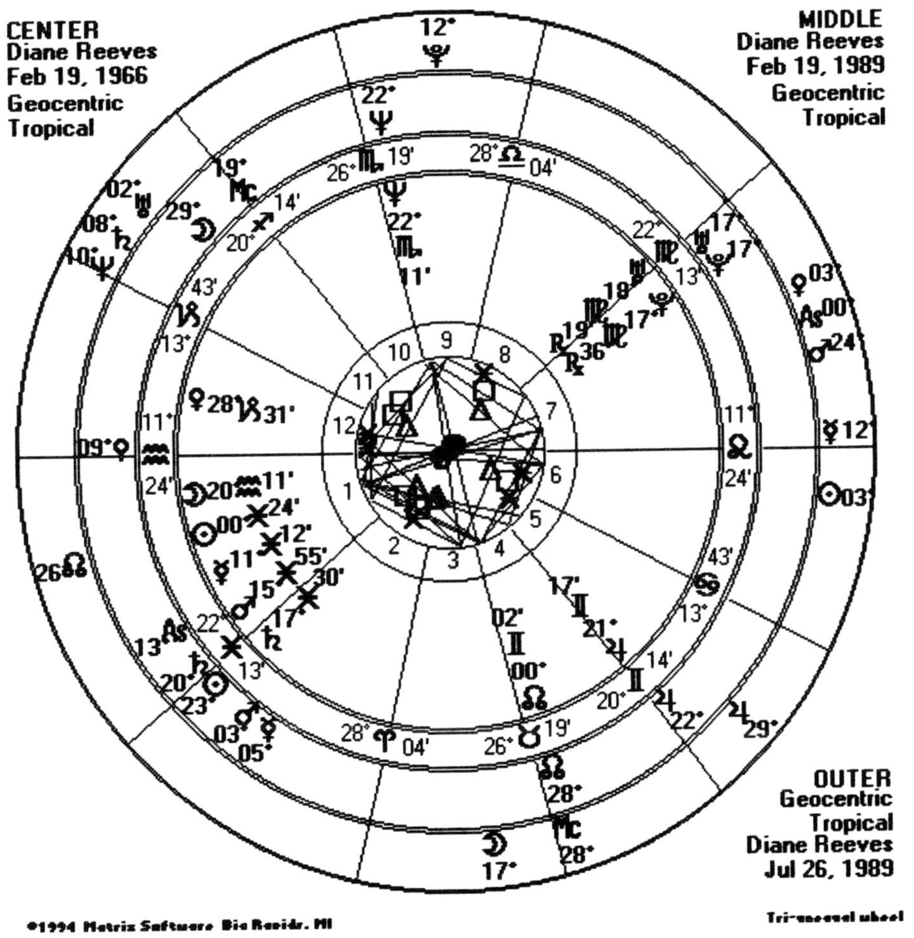

Chart 9 - Diane Reeves's Three-ring Chart with Natal, Progressed Chart for 1989 and Transits for July 26, 1989

Reeves' Natal Chart

Looking at Reeves' natal chart for the aspects with the closest orbs, we find Saturn at 17:30 Pisces in the first house of self opposing Pluto at 17:36 R Virgo and Uranus at 18:19 R Virgo both in the seventh house of one-to-one relationships. Saturn and Pluto are at critical degrees. These placements suggest a need to carefully evaluate relationships with others.

The Annular Solar Eclipse on September 11, 1988 took place at 18:41 Virgo and conjuncts Saturn and Uranus, indicating changes in these one-to-one relationships. Another indication of this is the Sun at 00:24 Pisces in the first house squares the North Node at 00:02 Gemini R in the fifth house love relationships.

A Partial Solar Eclipse on March 07, 1989 occurred at 17:09 Pisces. The eclipse falls in Reeves's first house and conjunct Saturn at 17:30 Pisces and indicates changes in personal affairs.

Reeve's Progressed Chart for 1989

In Reeves's progressed chart, we find progressed Uranus at the critical degree of 17:20 Virgo in the natal seventh house of one-to-one relationships. It conjuncts progressed Neptune at 17:00 Virgo, ruler of relationships, and opposes natal Mars at 15:55 Pisces, natal Saturn at the critical degree of 17:30 Pisces, and progressed East Point at 17:26 Pisces all in the natal first house of self. The progressed Moon at the culmination degree of 29:08 Sagittarius in the natal eleventh house of hopes and wishes opposes transiting Jupiter at 29:04 Gemini in the natal third house ruling the everyday environment. Progressed Venus is found at the critical degree of 09:39 Aquarius in the natal twelfth house of self-undoing and secret enemies.

The Yods

Second House	Progressed Sun	23:28 Pisces
Seventh House	Transiting Mars	24:56 Leo
Twelfth House	Natal Venus	28:31 Capricorn
Second House	Progressed Mars	03:51 Aries
Seventh House	Transiting Venus	02:57 Virgo
Twelfth House	Natal Part of Fortune	01:10 Aquarius

Astro*Carto*Graphy

Reeves's local space map for July 26, 1989, the day she disappeared, shows a Jupiter zone which includes Ferndale, WA and where Reeves lived. The Uranus zone focuses on a location in the coastal area of Washington. Uranus zones bring unexpected and sudden changes in one's circumstances.

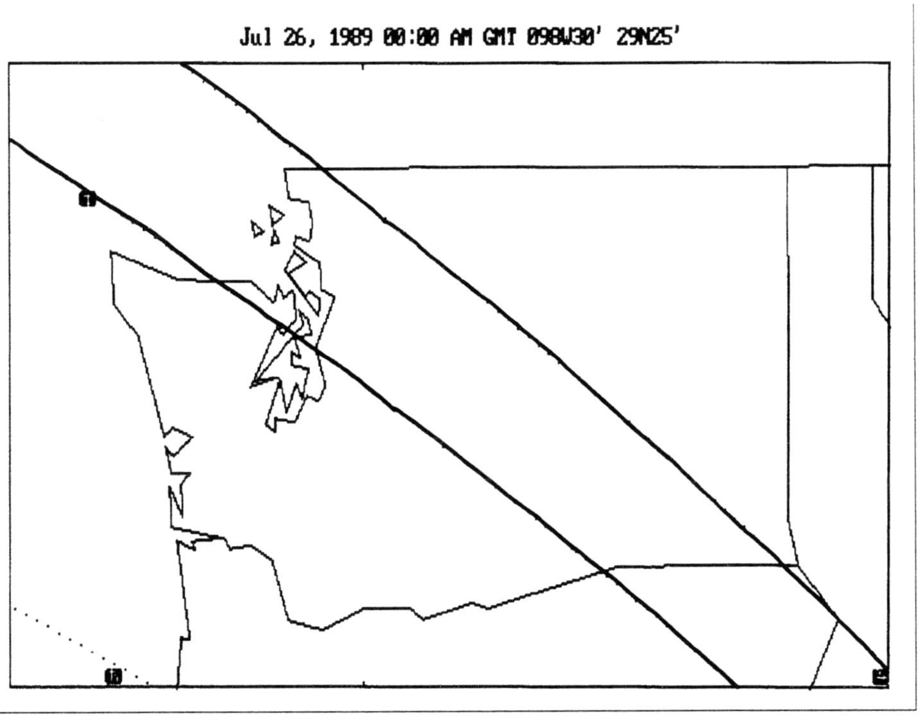

Map 2 - Diane Reeves's Local Space Map for July 26, 1989

Reeves's boyfriend's natal local space map shows a Mars zone which includes Reeves's birth place of San Antonio, TX. His North Node and the Moon zones focus on the same coastal area of the State of Washington as shown in Diane's local space map for the day that she disappeared.

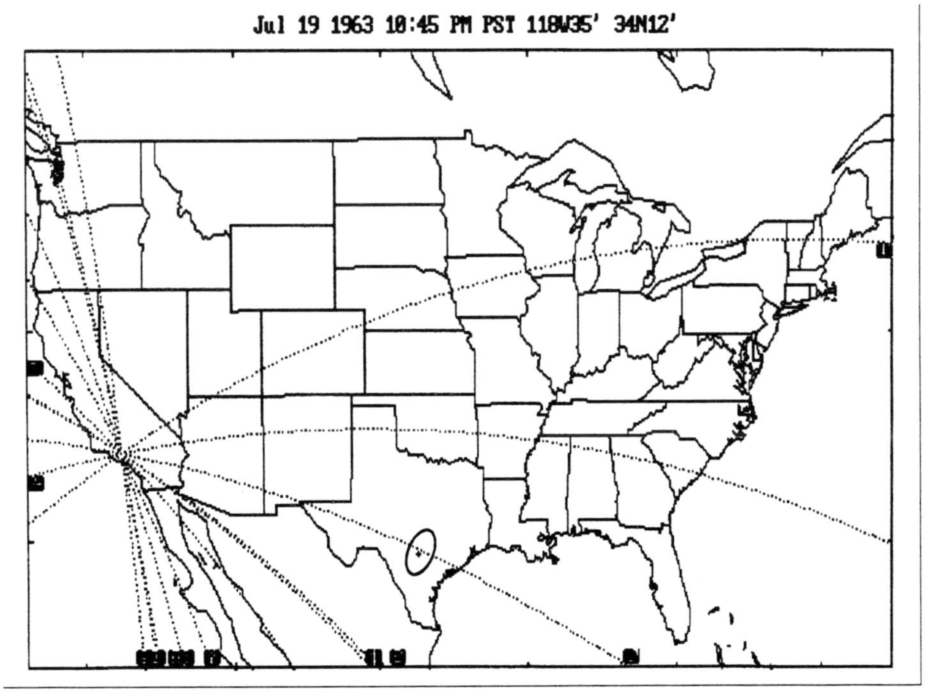

Map 3 - Reeves's Boyfriend's Natal Local Space Map for July 19, 1963

The emphasis of this coastal area is also indicated in both Reeves's boyfriend's progressed local space map for July 26, 1989 and his progressed Cyclo*Carto*Graphy map for July 26, 1989. The activated area is included under the North Node's zone in the local space map as well as the Mars zone in the progressed C*C*G map. Mars zones are indicative of areas where the typical behavioral patterns which are associated with this planet occur. Such behavioral patterns as aggressiveness and impulsiveness can in turn produce power struggles with others.

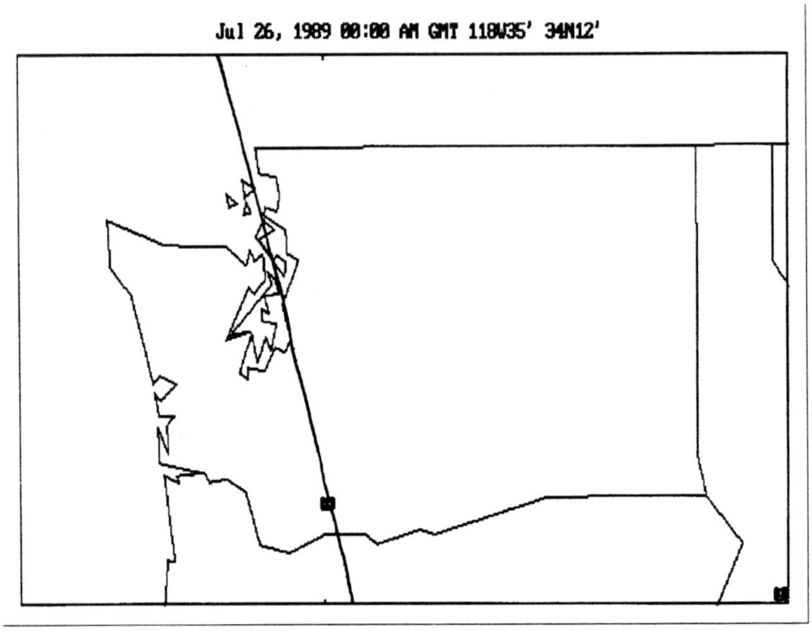

Map 4 - Reeves's Boyfriend's Local Space Map for July 26, 1989

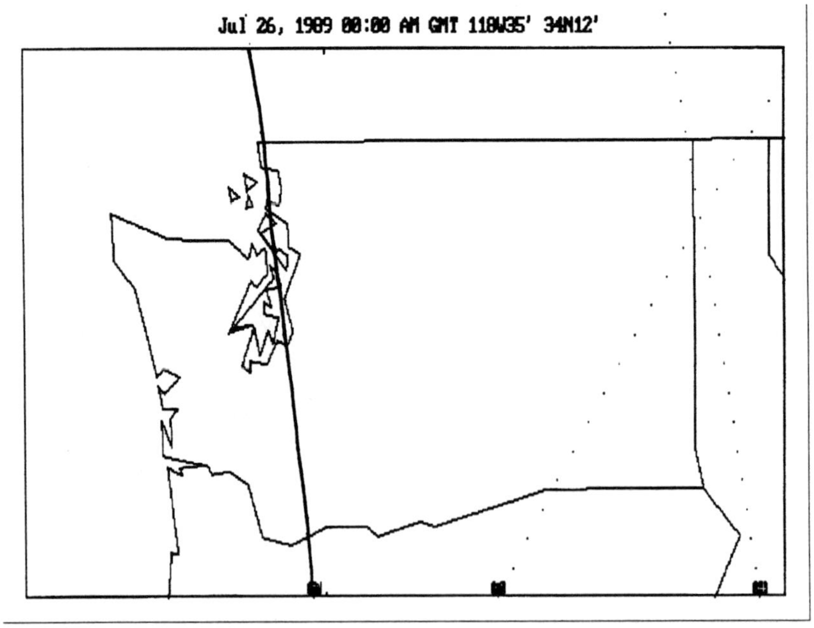

Map 5 - Reeves's Boyfriend's Progressed Cyclo*Carto*Graphy Map for July 26, 1989

In Conclusion

Reeves is still missing and the investigation into her disappearance remains open.

Birth Data for Diane Reeves

Birth date:	February 19, 1966		
Birth time:	06:14 AM/CST		
Birth place:	San Antonio, Texas		
Longitude:	098W29	Latitude:	29N26
Source:	Family		

Birth Data for Diane Reeves's Boyfriend

Birth date:	July 19, 1963		
Birth time:	11:45 PM/PDT		
Birth place:	Canoga Park, CA		
Longitude:	118W36	Latitude:	34N12
Source:	Reeve's family		

Placements and degrees of Diane Reeves' charts:
Reeves' Natal Placements

Moon	20::11 Aqu	01st	Neptune	22:11 Sco	09th
Sun	00:24 Pis	01st	Pluto	17:36 Vir R	07th
Mercury	11:12 Pis	01st	North Node	00:02 Gem R	04th
Venus	28:31 Cap	12th	Vertex	07:55 Vir	07th
Mars	15:55 Pis	01st	East Point	21:38 Aqu	01st
Jupiter	21:17 Gem	05th	Fortuna	01:10 Aqu	12th
Saturn	17:30 Pis	01st	Midheaven	26:19 Sco	
Uranus	18:19 Vir R	07th	Ascendant	11:24 Aqu	

Eleventh house position is 20:14 Sag
Twelfth house position is 13:43 Cap
Second house position is 22:13 Pis
Third house position is 28:04 Ari

Critical Degrees: Saturn, Pluto, and East Point
To be noted: Sun at 00 degrees Pisces and North Node at 00 degrees Gemini
Intercepted Signs: None

Progressed Placements for 1989

Moon	29:08 Sag	11th	Neptune	22:04 Sco R	09th
Sun	23:28 Pis	02nd	Pluto	17:00 Vir R	07th
Mercury	05:20 Ari R	02nd	North Node	28:49 Tau R	04th
Venus	09:39 Aqu	12th	Vertex	22:53 Vir	08th

Mars	03:51 Ari	02nd	East Point	17:27 Pis	01st
Jupiter	22:27 Gem	05th	Fortuna	19:09 Sag	10th
Saturn	20:19 Pis	01st	Midheaven	19:23 Sag	10th
Uranus	17:20 Vir R	07th	Ascendant	13:28 Pis	01st

Critical Degrees: Venus, Uranus, Pluto and East Point
To be noted: Moon at 29 degrees Sagittarius, Pluto at 00 degrees Virgo, and North Node approaching the ominous 29 degrees of Taurus

Transiting Placements for July 26, 1989

Moon	15:57 Tau	03rd	Uranus	02:09 Cap R	11th
Sun	03:29 Leo	06th	Neptune	10:23 Cap R	11th
Mercury	12:31 Leo	07th	Pluto	12:22 Sco	09th
Venus	02:57 Vir	07th	North Node	26:51 Aqu R	01st
Mars	24:56 Leo	07th	Vertex	13:32 Sag	10th
Jupiter	29:04 Gem	05th	East Point	27:08 Can	06th
Saturn	08:56 Cap R	11th	Fortuna	20:03 Tau	03rd

Critical Degrees: None
To be noted: Jupiter at 29 degrees Gemini, the turning point, in the fifth house of love relationships

Solar Return Placements for 1989 (Ferndale, WA)

Moon	13:01 Leo	12th	Neptune	11:36 Cap	05th
Sun	00:24 Pis	06th	Pluto	15:11 Sco R	03rd
Mercury	03:59 Aqu	05th	North Node	05:11 Pis R	07th
Venus	19:13 Aqu	06th	Vertex	05:59Aqu	05th
Mars	17:47 Tau	09th	East Point	23:19 Leo	12th
Jupiter	27:34 Tau	09th	Fortuna	17:40 Aqu	06th
Saturn	10:53 Cap	05th	Midheaven	27:54 Tau	
Uranus	04:19 Cap	04th	Ascendant	05:03 Vir	

Eleventh house position is 05:03 Can
Twelfth house position is 08:01 Leo
Second house position is 26:26 Vir
Third house position is 23:50 Lib

Critical Degrees: None
To be noted: Sun at 00 degrees Pisces
Intercepted Signs: Gemini in the tenth house of public recognition and Sagittarius in the fourth house ruling home and domestic environment

Placements and degrees of Reeves's Boyfriend's charts:
Natal Placements

Moon	19:21 Can	04th	Neptune	12:54 Sco R	07th
Sun	26:51 Can	04th	Pluto	10:24 Vir	06th
Mercury	04:09 Leo	05th	North Node	20:05 Can R	04th
Venus	15:41 Can	04th	Vertex	06:38 Lib	06th
Mars	25:53 Vir	06th	East Point	10:52 Ari	12th
Jupiter	18:49 Ari	01st	Fortuna	07:49 Ari	12th
Saturn	21:28 Aqu R	11th	Midheaven	09:10 Cap	
Uranus	03:16 Vir	05th	Ascendant	15:19 Ari	

Eleventh house position is 03:03 Aqu
Twelfth house position is 03:20 Pis
Second house position is 20:55 Tau
Third house position is 16:40 Gem

Critical Degrees: Sun at 26 Cancer in the fourth house ruling home and domestic environment and Saturn in the eleventh house of future hopes and wishes
To be noted: None
Intercepted Signs: None

Progressed Placements for 1989

Moon	01:14 Can	03rd	Neptune	13:00 Sco	07th
Sun	21:44 Leo	05th	Pluto	11:11 Vir	06th
Mercury	17:27 Vir	06th	North Node	18:42 Can R	04th
Venus	17:44 Leo	05th	Vertex	24:08 Lib	07th
Mars	11:47 Lib	06th	East Point	08:46 Tau	01st
Jupiter	19:26 Ari R	01st	Fortuna	01:13 Ari	12th
Saturn	19:36 Aqu R	11th	Midheaven	04:03 Aqu	11th
Uranus	04:46 Vir	06th	Ascendant	21:43 Tau	02nd

Critical Degrees: Sun in the fifth house of love relationships, Mercury, Uranus, Fortuna or Misfortuna
To be noted: None

Transiting Placements for July 26, 1989

Moon	26:48 Tau	02nd	Uranus	02:07 Cap R	09th
Sun	04:13 Leo	05th	Neptune	10:22 Cap R	10th
Mercury	14:01 Leo	05th	Pluto	12:22 Sco	07th
Venus	03:53 Vir	06th	North Node	26:49 Aqu R	11th
Mars	25:25 Leo	05th	Vertex	11:41 Lib	06th
Jupiter	29:14 Gem	03rd	East Point	19:03 Ari	01st
Saturn	08:53 Cap R	09th	Fortuna	19:08 Aqu	11th

Critical Degrees: Venus approaching 04 degrees Virgo
To be noted: Jupiter at 29 degrees in the third house of everyday environment

Solar Return Placements for 1989 (Ferndale, WA)

Moon	07:25 Aqu	07th	Neptune	10:34 Cap R	06th
Sun	26:51 Can	12th	Pluto	12:22 Sco R	04th
Mercury	28:16 Can	12th	North Node	27:13 Aqu R	08th
Venus	24:34 Leo	01st	Vertex	25:00 Sag	05th
Mars	20:34 Leo	01st	East Point	15:24 Can	12th
Jupiter	27:36 Gem	11th	Fortuna	17:28 Aqu	07th
Saturn	09:24 Cap R	06th	Midheaven	18:07 Ari	
Uranus	02:23 Cap R	05th	Ascendant	06:54 Leo	

Eleventh house position is 26:58 Tau
Twelfth house position is 05:53 Can

Second house position is 24:50 Leo
Third house position is 17:36 Vir

Critical Degrees: Sun
To be noted: None
Intercepted Signs: Gemini in the eleventh house of friends and future hopes and wishes and Sagittarius in the fifth house of love relationships

Monique Santiago
"Pebbles"

Monique Santiago, better known as "Pebbles," was last seen near her home in Albany, New York on March 29, 1990 at about 4:30 P.M. She was 11-1/2 years old at the time of her disappearance. Her mother reported her missing on March 30, 1990 at 12:15 AM. The longitude and latitude for Albany, NY are 73W47 and 42N39.

Santiago had moved to the Albany area about two years earlier with her mother, Anna Graniela, and her sister Lisa. They previously lived in Bronx, New York. At the time of Santiago's disappearance, her mother was living with a boyfriend who had recently been released from prison, and she was three months pregnant.

ALBANY POLICE DEPARTMENT
JUVENILE UNIT

NAME: SANTIAGO, Monique NMN
NICKNAME/AKA: "PEBBLES"
SEX: Female
Race: Hispanic / PR
DOB: 09/11/78
HGT: 4 feet 9 inches
WGT.: 90 lbs.
HAIR: Black, Long and curly
EYES: Brown
LSW: Blue Jeans, Turquoise Blouse, Black Shoes

NARRATIVE:
Last seen an Thursday March 29, 1990 at 4:30 PM.
Reported missing on Friday March 30, 1990 at 12:15 AM by the mother. Incident #90-23676

ALBANY, NEW YORK POLICE DEPARTMENT
JUVENILE UNIT 536 WESTERN AVE.
ALBANY, NEW YORK (518) 468-5666

Illustration 8 - Monique Santiago's Missing Person Flyer (Santiago)

Santiago was mature beyond her years in many ways. She was street smart, and the police said it had been reported that she was prostituting herself. Her mother said that she believed her daughter was kidnapped because she had no history of running away. However, the police found no evidence to support a kidnapping, but they did not discount that foul play was involved in her disappearance. The police also theorized that she may have run away to the New York City area. There were no substantial clues into Pebbles' disappearance.

Daughter, 11, missing since March 29 09/04

Albany mother's torment continue

By Joe Mahoney

No leads to missing

By Joe Mahoney 05/12/90

ALBANY — Six weeks after Monique "Pebbles" Santiago was reported missing, city police remain baffled in investigating the disappearance of the 11-year-old girl.

The girl is now being listed by police as missing under suspicious circumstances, though investigators have found no evidence pointing to or discounting the possibility of foul play.

"All the leads we had have checked out negative," said Albany Police Lt. Charles Geyer, who supervises the department's juvenile unit. "We've checked out everybody we could possibly check out."

ALBANY COUNTY

Albany Times Union

Search for girl enters 5th year 5/5/94

Age-enhanced photo helps police keep investigation alive

Missing Albany girl's case to be on TV show

By Joe Mahoney 12/26/90

Illustration 9 - Albany Times Union (Santiago)

A liturgy for Monique Santiago was held on September 15, 1990. The following month, several new leads on her disappearance were received by police when her case was aired on "America's Most Wanted." On February 6, 1991, Santiago's mother, then named Anna Castro, 31, and eight months pregnant, was arrested and charged with heroin possession, cocaine possession, conspiracy and criminal facilitation, along with five other charges.

Santiago's Natal Chart

In Santiago's natal chart, there are no critical degrees, but there is one yod. This yod is formed by Venus at 03:40 Scorpio in the third house of everyday environment sextiling the Moon at 04:56 Capricorn in the fourth house of home and personal areas. These two planets form inconjuncts to the Midheaven at 02:57 Gemini.

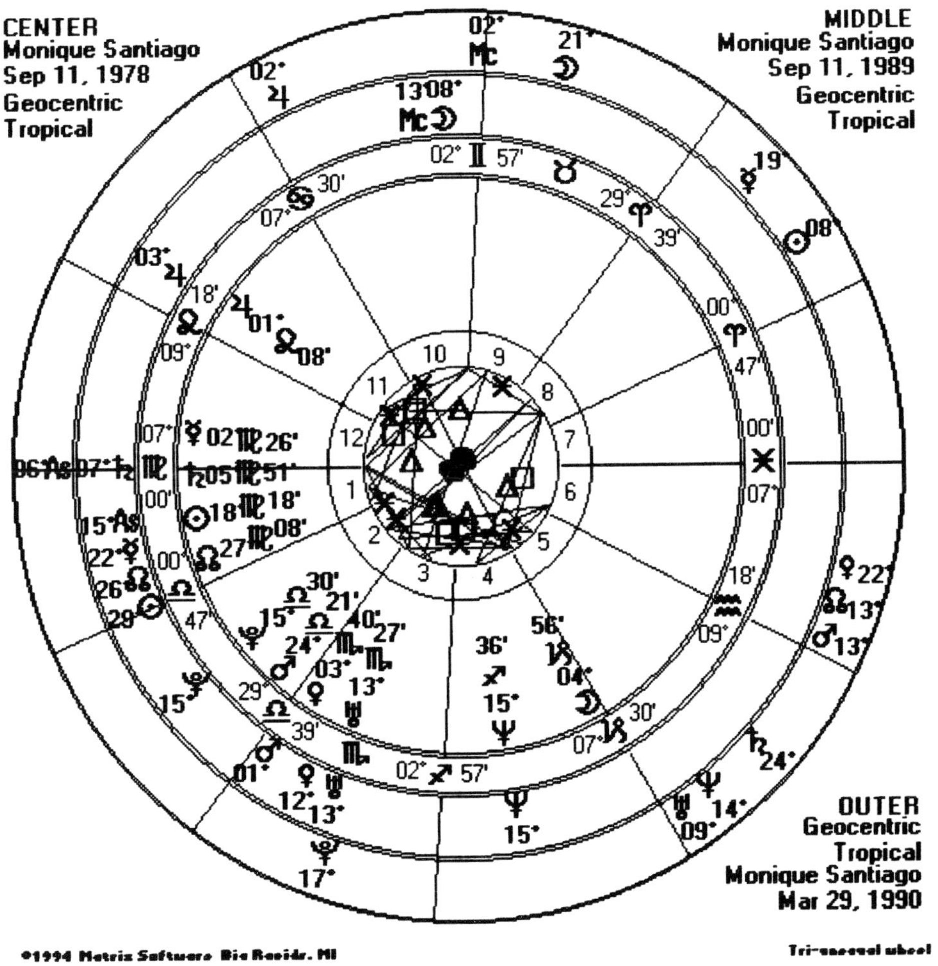

Chart 10 - Monique Santiago's Three-ring Chart with Natal, Progressed Chart for 1989 and Transits for March 29, 1990

The Progressed Chart for 1989

The progressed Sun is at 29:02 Virgo in the natal first house of self suggesting that during that year Santiago needed to use care in the way she expressed herself to others. Progressed Venus at 12:20 Scorpio and Uranus at 13:57 Scorpio both in the natal third house opposes the progressed Midheaven at 13:41 Gemini in the tenth house of public recognition and reputation.

Transiting Aspects for March 29, 1990

There are trends in Santiago's transiting planets that indicate danger to her and even physical harm at the time of her disappearance. The following transits are taken in part from *Transits* by Robert Ebertin:

Mer inc Sun -- S 01:35 A poor atmosphere for negotiations. Obscurity, absent-mindedness, nervousness and a disinclination to work.

Mar sqr Ura -- S 00:06 A tendency of a contradictory personality, irritation and inner turmoil. A desire to release inner tension and to stand a test of nerves. A tendency to suffer accidents and injury. Danger through machinery and modern means of transportation.

Jup sxt Mer -- S 00:06 Good common sense, wealth of plans, organizational talent and optimism. Can attain success as merchant and advantageous transactions. Successful trips.

Jup tri Ven -- A 01:08 A joy for love and sincerity. A feeling of attractiveness making one popular, receipt of affection and harmonious relations with others. Happy friendships.

Sat sqr Mars -- A 00:03 Dissatisfaction or displeasure stemming from great difficulties and antagonisms which involve severity, harshness, insubmissiveness, weakness and inclination to illness. Danger of injury or mishaps, mental and emotional pain, and sorrow and harm from others.

Nep sxt Ura -- S 01.02 New problems, new ideas and views. Possible inspiration, broadening of perspective, peculiar mental states and travel. Indication of foreign matters, turning points of life and a relationship to the occult.

Nep sqr Plu -- A 01:01 Peculiar mental states and possession by fantastic ideas. A collapse of an idea, self-torture, heavy losses and possible strange illnesses.

Plu sxt Sun -- S 00:56 A realization of new ideas.

Node sqr Ura -- S 00:21 Sudden incidents in communal life, upsets with others or conflicts.

Node sxt Nep -- A 01:48 Great expectations from a cooperative effort. Secret relationships.

Node tri Plu -- A 01:42 Exposure to relationships with possible fateful turns.

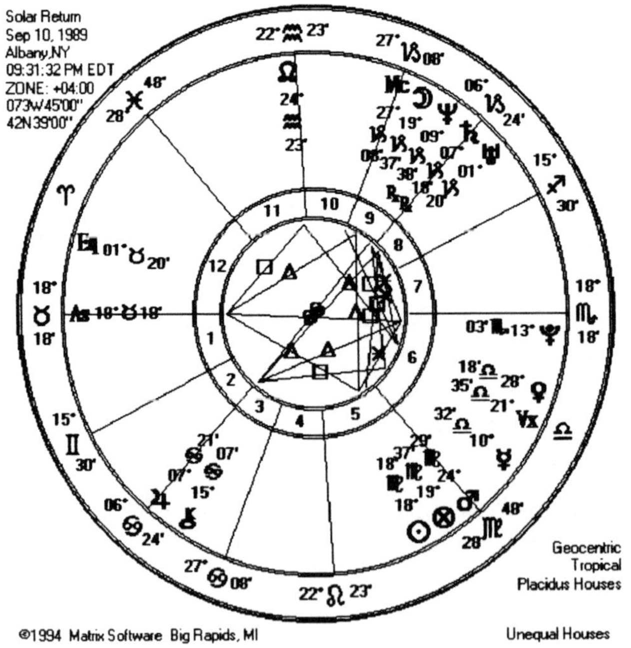

Chart 11 - Monique Santiago's Solar Return Chart for 1989

An Annular Solar Eclipse occurred on January 26, 1990 occurred at 06:34 Aquarius. The eclipse falls in Santiago's fifth house and indicates changes in children, schools, pleasures and can involve risk taking.

The Solar Return Chart for 1989 (Albany, NY)

There are no critical degrees in Santiago's solar return chart. However, Libra is intercepted in the sixth house of health and Aries in the twelfth house of karma or self-undoing. Jupiter at 07:21 degrees of Cancer in the third house of everyday environment is in close opposition to Saturn and Neptune both in the ninth house of abstract mind. A grand trine is formed by the Ascendant at 18:18 Taurus, the Sun at 18:18 Virgo and Part of Fortuna or Misfortuna at 19:37 Virgo both in the fifth house of love relationships and the Moon at 19:37 Capricorn in the ninth house.

The Yods or Fingers of Fate

There are two yods in Santiago's combined charts. One is comprised of progressed Venus at 12:20 Scorpio and progressed Uranus at 13:57 Scorpio both in the natal third house of everyday environment. These planets sextile transiting Neptune at 14:29 Capricorn in the natal fifth house of love relationships and risk taking. Progressed Venus, progressed Uranus and transiting Neptune inconjunct progressed Midheaven at 13:41 Gemini in the natal tenth house. Santiago's unexpected disappearance from her everyday environment, which is thought to be outside and in front of her home, is influenced by progressed Uranus' participation in this yod and placement in the third house. Transiting Neptune in the fifth house created confusion and illusions in all fifth house matters.

Santiago's other yod is comprised of natal Mars at 24:21 Libra in the natal second house of finances, the natal Part of Fortuna or Misfortuna at 23:38 Sagittarius in the natal fourth house home and personal matters and the transiting Moon at the critical degree of 21:52 Taurus in the ninth house of abstract mind.

Yod No. 1

Third House	Progressed Venus	12:20 Scorpio
Third House	Progressed Uranus	13:57 Scorpio
Fifth House	Transiting Neptune	14:29 Capricorn
Tenth House	Progressed Midheaven	13:41 Gemini

Third House, Progressed Venus/Uranus in Scorpio

What follows is an interpretation of the yod in Santiago's charts using repetitive Basic Keywords from *The Rulership Book* by Rex E. Bills.

Third House

Automobiles, trips by	Everyday environment	Persons sharing home,

Travel, local
Railroads, rail yards
Parent, parent-in-law, step-parent
unrelated

Progressed Venus
Children, in general
Friends, friendships
Immorality
Immaturity
Love affairs, love-making, lovers
Sexual intercourse, sexual relations
Motels
Offers, monetary
Venereal disease

Progressed Uranus
Automobiles
Chances, the laws of chance
Changes, sudden
Deaths, sudden
Perverts, perversion
Railroads, railroad yards
Relatives, in general
Runaways
Strangers
Trucks, truck drivers
Unexpected, the
Upsets, general
Vans

Scorpio
Dead, matters of the dead
Rubbish, dumps
Venereal disease
Sexual intercourse

Fifth House, Transiting Neptune in Capricorn

Fifth House
Children, in general
Parent, step-parent
Love affairs, love-making, lovers
Pregnancy
Promiscuity
Sexual intercourse

Transiting Neptune
Abnormal behavior
Addicts
Crime, crimes, criminals
Drugs, in general
Exotica
Kidnapping, kidnappers
Narcotics
Occultism
Perverts, perversion

Capricorn
Abandoned places
Engineers, in general

Tenth House, Progressed Midheaven in Gemini

Tenth House
Aunts/uncles, mother's side
Children, in general
Fate, one's
Parent, step-parent

Gemini
Automobiles
Travel, local
Railroads, transport, yards
Journeys, short

Yod No. 2

Second House	Natal Mars	24:21 Libra
Fourth House	Natal Part of Fortuna	23:30 Sagittarius
Ninth House	Transiting Moon	21:52 Taurus

Second House, Natal Mars in Libra

Second House
Children, in general
Incomes
Money
Parent, step-parent

Natal Mars
Ability, one's working
Anger
Arguments
Torture
Crime, crimes, criminals
Knives
Crimes of violence
Death by accident/violence
Deaths, sudden
Engineers, in general
Brutality
Injuries by violence
ale relatives
Danger, dangerous occupations
Males, in general
Murder
Sexual intercourse
Gunshot wounds
Violence, crimes of violence

Libra
Love, lovers
Unions

Fourth House, Natal Part of Fortuna/Misfortuna, Sagittarius

Fourth House
Burial, one's
Death, manner of one's
Environment
Families, one's family
Home, one's home life
Journeys, short
Love affairs, sorrow through, losses in
Motels
Mother, one's parent
Parent, step-parent
People, elderly

Sagittarius
Foreign people

Ninth House, Transiting Moon, Taurus

Ninth House
Children, in general
Foreign people
Religion, rituals

Transiting Moon
Environment
Homes, one's home life
Motels
Mother, one's
Occultists
Pregnancy

Taurus

Money, one's earning capacity	Strangulation	Suffocation

In Conclusion

Santiago is still missing and the investigation into her disappearance remains open. A May 5, 1994 article in the *Albany Times Union* reported that the search for her has entered its fifth year without any substantial clues. The police have also age-enhanced a photograph of Santiago to show her at the age of 16. It was reported in the same article that Santiago's mother, Anna Castro, lost her bout with cancer the year before, without ever knowing what had happened to her daughter.

Illustration 10 - Age-enhanced photo of Santiago

Birth Data for Monique Santiago

Birth date:	September 11, 1978		
Birth time:	05:39 AM/EST		
Birth place:	Bronx, NY		
Longitude:	073W54	Latitude:	40N51
Source:	Her mother		

Placements and degrees of Santiago's charts:
Natal Placements

Moon	04:56 Cap	04th	Neptune	15:36 Sag	04th
Sun	18:18 Vir	01st	Pluto	15:30 Lib	02nd
Mercury	02:26 Vir	12th	North Node	27:08 Vir R	01st
Venus	03:40 Sco	03rd	Vertex	05:06 Aqu	05th
Mars	24:21 Lib	02nd	East Point	28:46 Leo	12th
Jupiter	01:08 Leo	11th	Fortuna	23:38 Sag	04th
Saturn	05:51 Vir	12th	Midheaven	02:57 Gem	
Uranus	13:27 Sco	03rd	Ascendant	07:00 Vir	

Eleventh house position is 07:30 Can
Twelfth house position is 09:18 Leo
Second house position is 00:47 Lib
Third house position is 29:39 Lib

Critical Degrees: None
To be noted: East Point approaching 29 degrees of Leo
Intercepted Signs: Taurus in the ninth house of abstract mind and Scorpio in the third house of everyday environment

Progressed Placements for 1989

Moon	08:44 Gem	10th	Neptune	15:43 Sag	04th
Sun	29:02 Vir	01st	Pluto	15:54 Lib	02nd
Mercury	22:05 Vir	01st	North Node	26:33 Vir R	01st
Venus	12:20 Sco	03rd	Vertex	23:43 Aqu	06th
Mars	01:42 Sco	03rd	East Point	10:49 Vir	01st
Jupiter	03:04 Leo	11th	Fortuna	25:41 Tau	09th
Saturn	07:12 Vir	01st	Midheaven	13:41 Gem	10th
Uranus	13:57 Sco	03rd	Ascendant	15:59 Vir	01st

Critical Degrees: None
To be noted: Sun at 29 degrees of Virgo in the first house of self

Transiting Placements for March 29, 1990 (4:30 PM/EST)

Moon	21:52 Tau	09th	Uranus	09:30 Cap	05th
Sun	08:56 Ari	08th	Neptune	14:29 Cap	05th
Mercury	19:53 Ari	08th	Pluto	17:22 Sco R	03rd
Venus	22:28 Aqu	06th	North Node	13:48 Aqu R	06th
Mars	13:33 Aqu	06th	Vertex	29:55 Vir	06th
Jupiter	02:32 Can	10th	East Point	14:23 Vir	01st
Saturn	24:18 Cap	05th	Fortuna	01:33 Sco	03rd

Critical Degrees: Moon
To be noted: Vertex at 29 degrees of Virgo

Solar Return Placements for 1989 (Albany, NY)

Moon	19:37 Cap	09th	Neptune	09:38 Cap R	09th
Sun	18:18 Vir	05th	Pluto	13:03 Sco	06th
Mercury	10:32 Lib	06th	North Node	24:23 Aqu R	11th
Venus	28:18 Lib	06th	Vertex	21:35 Lib	06th
Mars	24:29 Vir	05th	East Point	01:20 Tau	12th
Jupiter	07:21 Can	03rd	Fortuna	19:37 Vir	05th
Saturn	07:18 Cap R	09th	Midheaven	27:08 Cap	
Uranus	01:20 Cap	08th	Ascendant	18:18 Tau	

Eleventh house position is 22:23 Aqu
Twelfth house position is 28:48 Pis
Second house position is 15:30 Gem
Third house position is 06:24 Can

Critical Degrees: None
To be noted: None
Intercepted Signs: Aries in the twelfth house of self-undoing and Libra in the sixth house of health and service

Agnes Shoe
A Case Remembered

More than twenty years after Agnes Shoe's "suspicious" disappearance, a police investigator still remembers this unsolved case. Agnes Shoe had red hair and blue eyes, and she was to have taken a trip to New York City on June 18, 1979. However, according to her husband who was out of town at the time, his wife never reached her destination. Her clothes and car were still at her home in Glenville, New York. The longitude and latitude for Glenville, NY are 73W50 and 41N04.

Missing woman case not forgotten

Detective still searching after 17 years

[illegible]

Searchers stumped

[illegible]

See DETECTIVE, Page B6

Illustration 11 - Schenectady Gazette November 10, 1996 (Shoe)

Investigators found no sign of foul play or any note written by Shoe. There was an extensive manhunt after her disappearance that combed the woods, fields and neighborhood surrounding her home. Nothing was found. Relatives and friends said it was not like Shoe to take off and not to contact them. In fact, no one has ever heard from her again.

Shoe had been a Franciscan nun who lived in Connecticut, but left the order, and moved to New Jersey. She met Richard Shoe when they worked at the Six Flag Amusement Park and they later married. It was her first marriage and his second. After their marriage, they moved to the Capital District Region of upstate New York.

Agnes Shoe's Natal Chart

Shoe's natal Moon is at the critical degree of 00:08 Libra in the third house ruling her everyday environment. Her Ascendant is within two minutes of the 29th degree of Cancer. Cancer rules the fourth house of home and personal affairs.

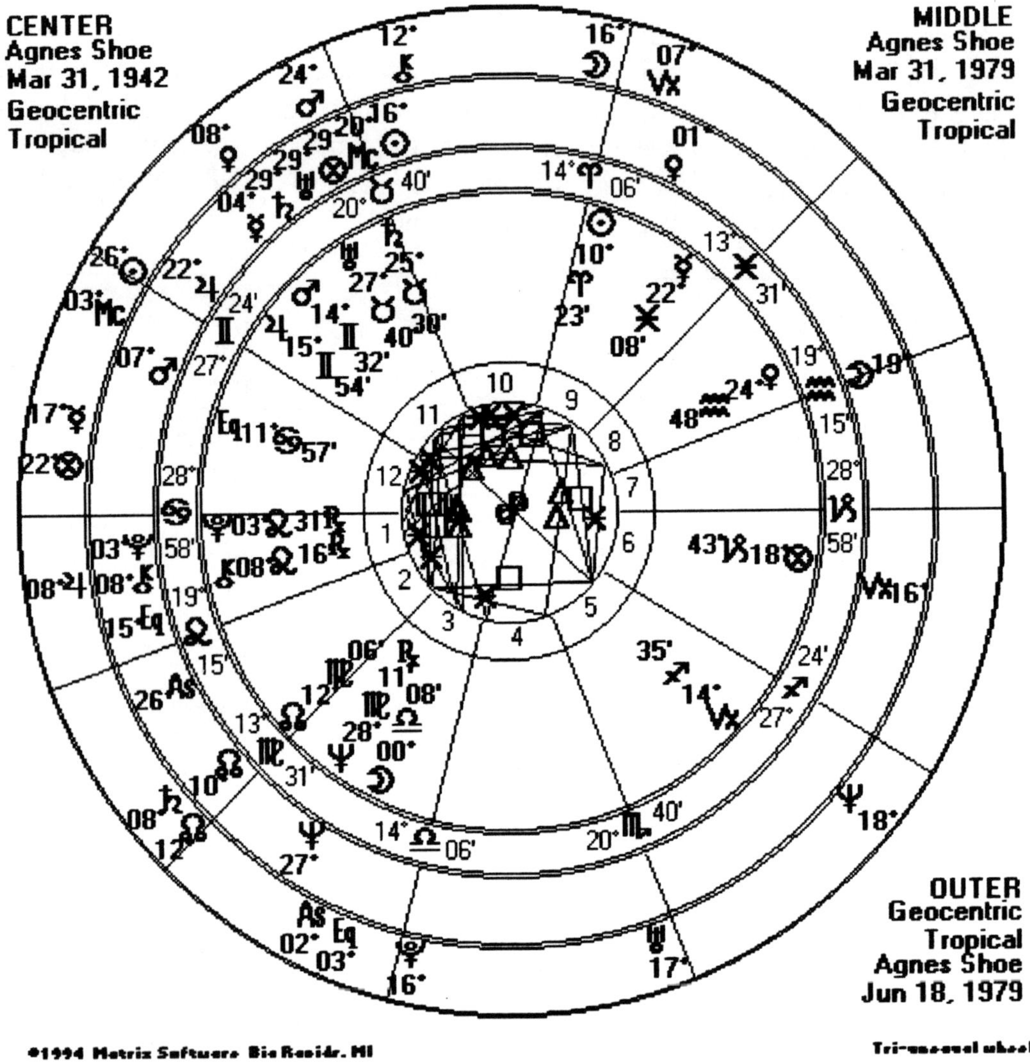

Chart 12 - Agnes Shoe's Three-ring Chart with Natal, Progressed Chart for 1979 and Transits for June 18, 1979

Looking at Shoe's natal chart, we find the Moon at the critical degree of 00:08 Libra in polarity with Aries which rules injury to the head. The outer planet of Jupiter at 15.54 degrees of Gemini is conjunct Mars at 14:32 degrees of Gemini both in the eleventh house of friends, hopes and wishes. Uranus at 27:40 degrees of Taurus in the eleventh house trines Neptune at 28:11 Virgo R in the third house ruling the everyday environment. Neptune also sextiles the Ascendant at 28:58 Cancer. Venus at 24:48 degrees of Aquarius in the eighth house squares Saturn at 25:30 degrees of Taurus in the eleventh house.

The Fixed Star Algol at 25:03 Taurus conjuncts natal Saturn both in the eleventh house and squares Venus at 24:48 Venus in the eighth house. Algol has a Saturn, Mars, Uranus, and Pluto influence. It is said to be the most violent star, and it is connected with strangulation, beheading or other violent injury to the head, throat and neck.

The Progressed Chart for 1979

In Shoe's progressed chart Saturn at the ominous 29:52 Taurus conjuncts both Uranus at 29:34 Taurus and Fortuna or Misfortuna at 29:24 Taurus all in the natal eleventh house of friends, hopes and wishes. These placements trine progressed Neptune at 27:21 Virgo R in the natal third house of everyday environment. Progressed Mercury is at the critical degree of 04:28 Gemini in the natal eleventh house and within an allowable orb of the 29 degree conjunction of Saturn, Uranus and Fortuna.

Shoe's Transits Interpreted for June 18, 1979 - 1:00 PM/EST

Let's look at interpretations in-part from *Predictive Astrology* by Frances Sakoian and Louis Acker and *Modern Transits* by Lois Rodden. The degrees and minutes of separation are listed for each aspect along with whether it is a separating (S) or applying (A) aspect.

Moon Sxt Jup -- S 00:53 Ideas for incorporating educational and cultural values into the family life characterize this transit. There is an increased prosperity in family finances and businesses.

Sun Sqr Nep -- A 01:17 Problems arising from hypocrisy, extravagance and an inflated sense of self-importance characterizes this transit. There are impractical attitudes. *** It's a vaguely low-tone day.

Ven Sxt Sun -- A 01:53 Enjoyment of social activities. A favorable day for short-pleasure trips. *** Social affairs may have a divided interest.

Mars Sqr Ven -- A 00:04 Problems in romantic and sexual relationships characterize this transit. Jealousy is often expressed in business, marital or romantic relationships by either the individual or their associates. Sexual overindulgence and lack of consideration for the feelings and moods of others are among the dangers of this transit. Joint finances, insurance, taxes, alimony and goods of the dead are often the source of difficulties. *** A physical outlet is satisfactory to restless feelings.

Mars Con Sat -- A 00:46 Hard physical work, frustrated impulses to action, and suppressed anger and resentment characterize this transit. *** Trauma or stress is in your environment; maintain control and responsible action.

Sat Inc Sun -- A 01:57 Frustration of personal ambition and need for self-expression characterize this transit. This is an unfavorable transit in general for social and romantic activity. *** Varied pressures and worries are a strain.

Ura Inc Jup -- S 01:36 Increased tolerance for all points of view and ways of life characterize this transit. *** Sudden changes exact a heavy toll.

Plu Tri Mars -- S 01:54 Intensified energy and will power and the exercise of personal leadership in business and professional affairs characterize this transit. This transit occurs once in the average life span. *** Under stress, competition and repression you can transmute your passion into combative action.

Plu Tri Jup -- S 00:32 Interest in spiritual regeneration and self-improvement characterizes this transit. Benefit may come to the individual through religion, higher education and occult or esoteric doctrines and studies. This aspect occurs only once in the average life span. ** This can be a turning point, during which you move through a repression or crisis in your next step of growth.

Eclipses

The Partial Solar Eclipse of October 2, 1978 occurred at 08:43 Libra which took place in the third house. It affects the areas of: One's habits, everyday environment, short-distance journeys and siblings. The next Total Solar Eclipse of February 26, 1979 occurred at 07:31 Pisces and took place in the eighth house. It affects the areas of: Other's stuff, resources shared with other people, sexual attitudes and an interest in life after death.

The Solar Return Chart for 1979 (Glenville, NY)

Jupiter at 29:03 Cancer in the first house of self sextiles the Moon at 27:03 Taurus in the eleventh house of friends, hopes and wishes. Jupiter also trines Mercury at 27:41 Pisces R in the ninth house of the abstract thought. The Sun at 10:23 Aries in the tenth house of public recognition inconjuncts Saturn at 08:20 Virgo R and Fortuna at 12:28 Virgo both in the third house of everyday environment.

Pluto at 18:04 Libra R in the fourth house of home and personal matters inconjuncts the South Node at 16:29 Pisces R in the ninth house of abstract thought. Lastly, Mars at 25:03 Pisces in the ninth house trines Shoe's Ascendant at 25:48 Cancer.

The Yod or Fingers of Fate

There is a yod formed between Shoe's natal and solar return charts. It is formed by natal Pluto at 03:31 Leo R in the natal first house of self, natal Moon at the critical degree 00:08 Libra in the natal third house everyday environment and solar Venus at 03:03 Pisces in the natal eighth house of shared resources, death and regeneration.

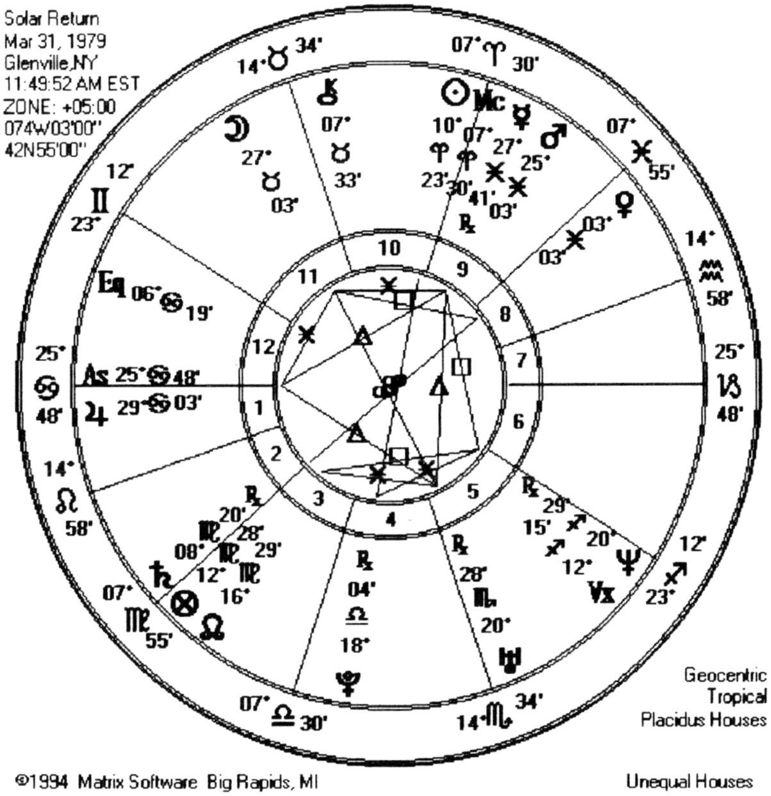

Chart 13 - Agnes Shoe's Solar Return Chart for 1979

The Yod

First House	Natal Pluto	03:31 Leo R
Third House	Natal Moon	00:08 Libra
Eighth House	Solar Venus	03:04 Pisces

The yod includes the first house of self, third house of everyday environment and eighth house of shared resources. The house keywords from *Astrological Thesaurus* by Michael Munkasey interprets these houses as follows:

First House	Actions upon one's life	Anger
	Danger, peril	Aggression, assault
	Beat, beating	Bullet
	Callous actions	Firearm
	Gunshot	Rash action
	Violence	Home front
Third House	Cleverness	Neighborhood
	Alias	Alibi
	Cheating	Daily comings and goings
	Deceit	

Eighth House	Cause of death	Jointly held property
	Partner's money	Accidental death
	Alimony	Compulsive acts
	Disappearance, vanishing	Dismemberment
	End, death	Execution, slaying
	Finances of spouse or partner	Gain due to death
	Infidelity	Money of a deceased person
	Murder	Property jointly held
	Violent crime	Cesspool

Picture 5 - Agnes Shoe

In Conclusion

Agnes Shoe is still missing and the investigation into her disappearance remains open.

Birth Data for Agnes Shoe

Birth date:	March 31, 1942		
Birth time:	01:50 PM/EWT		
Birth place:	Columbus, OH		
Longitude:	083W00	Latitude:	39N58
Source:	Police investigator		

Placements and degrees of Shoe's charts:
Natal Placements

Moon	00:08 Lib	03rd	Neptune	28:11 Vir R	03rd
Sun	10:23 Ari	09th	Pluto	03:31 Leo R	01st
Mercury	22:08 Pis	09th	North Node	12:06 Vir R	02nd
Venus	24:48 Aqu	08th	Vertex	14:35 Sag	05th
Mars	14:32 Gem	11th	East Point	11:57 Can	12th
Jupiter	15:54 Gem	11th	Fortuna	18:43 Cap	06th
Saturn	25:30 Tau	11th	Midheaven	14:06 Ari	
Uranus	27:40 Tau	11th	Ascendant	28:58 Can	

Eleventh house position is 20:40 Tau
Twelfth house position is 27:24 Gem
Second house position is 19:15 Leo
Third house position is 13:31 Vir

Critical Degrees: Moon
To be noted: Ascendant approaching the 29th degree of Cancer
Intercepted Signs: None

Progressed Placements for 1979

Moon	19:28 Aqu	08th	Neptune	27:21 Vir	03rd
Sun	16:31 Tau	10th	Pluto	03:34 Leo	01st
Mercury	04:28 Gem	11th	North Node	10:08 Vir R	02nd
Venus	01:47 Ari	09th	Midheaven	20:15 Tau	10th
Mars	07:00 Can	12th	Ascendant	26:27 Leo	02nd
Jupiter	22:38 Gem	11th	Vertex	16:57 Cap	06th
Saturn	29:52 Tau	11th	East Point	15:21 Leo	01st
Uranus	29:34 Tau	11th	Fortuna	29:24 Tau	11th

Critical Degrees: Mercury
To be noted: Saturn, Uranus and Fortuna and Misfortuna at the ominous degree of 29 Taurus

Transiting Placements for June 18, 1979 - 01:00 P.M./ EST

Moon	16:47Ari	10th	Uranus	17:30 Sco R	04th
Sun	26:54 Gem	11th	Neptune	18:54 Sag R	05th
Mercury	17:21 Can	12th	Pluto	16:26 Lib R	04th
Venus	08:30 Gem	11th	North Node	12:17 Vir R	02nd
Mars	24:44 Tau	11th	Vertex	07:39 Ari	09th
Jupiter	08:23 Leo	01st	East Point	03:42 Lib	03rd
Saturn	08:26 Vir	02nd	Fortuna	22:35 Can	12th

Critical Degrees: None

To be noted: None

Solar Return Placements for 1979 (Glenville, NY)

Moon	27:03 Tau	11th	Neptune	20:29 Sag R	05th
Sun	10:23 Ari	10th	Pluto	18:04 Sco R	04th
Mercury	27:41 Pis R	09th	North Node	16:29 Vir R	03rd
Venus	03:03 Pis	08th	Vertex	12:15 Sag	05th
Mars	25:03 Pis	09th	East Point	06:19 Can	12th
Jupiter	29:03 Can	01st	Fortuna	12:28 Vir	03rd
Saturn	08:20 Vir R	03rd	Midheaven	07:30 Ari	
Uranus	20:28 Sco R	05th	Ascendant	25:48 Can	

Eleventh house position is 14:34 Tau
Twelfth house position is 23:12 Gem

Second house position is 14:58 Leo
Third house position is 07:55 Vir

Critical Degrees: None
To be noted: Jupiter at 29 degrees Cancer and the Ascendant approaching the critical degree of 26 Cancer.
Intercepted Signs: None

ABDUCTED AND MURDERED CHILDREN

Every year, thousands of children disappear. Many of them are abducted by a stranger or a parent while others run away from home. In today's society, there are also those children who are known as "throwaways." These are the children whose parents no longer want them around and put them out of their homes. The statistics on missing children are alarming.

James Tedesco and Michele Paludi in the Introduction to their book *Missing Children* state that "Although generalizations are difficult given some problems in identifying children as 'missing' or 'abducted,' the best estimate is that child abduction is the most widespread of all forms of child victimization studied.

In 1992, a total of 27,553 cases of missing children were reported in New York State through the Missing Children Register. The majority of the missing children cases involved suspected runaways..."

Some of these missing children return home, but most do not. Many parents never give up in the search for their missing child and live with the constant pain of not knowing whether their child is alive or dead. Often the missing children are murdered. When, and if, their bodies are found, it's as if the abductors simply wanted to dispose of the children's bodies like items they no longer had any use of.

Two widely publicized cases of missing children in 1993 were that of Holly Piirainen of Massachusetts and Polly Klaas of California. Their names were linked in newspapers from coast to coast while their parents appealed to the country to find their daughters and, then, their daughter's killers.

Holly Kristin Piirainen
1983 - 1993

Holly Piirainen was out walking near the home of her paternal grandparents in Sturbridge, Massachusetts when she disappeared on Thursday, August 5, 1993 at about 12:30 PM/EDT. The longitude and latitude for Sturbridge, MA are 072W05 and 42N06. When her two younger brothers were told to go to a neighbor's house to get their sister, the boys found one of Piirainen's white sneakers lying by the road. She was reported missing about 20 minutes later.

Hundreds of people volunteered to search the area where Piirainen was last seen, but they found no trace of her. Others organized fund raisers and distributed posters with Piirainen's picture. On August 10, her story appeared on "America's Most Wanted."

Richard Piirainen and his former wife, Christina, had joint custody of the blond, blue-eyed girl and they were both devastated by her disappearance. Even though there was little evidence, they never gave up hope that she would be found alive. However, on October 23, 1993, eleven weeks after her disappearance, Piirainen's skeletal remains were found by hunters in Brimfield, MA. The area had not been searched and was located five or six miles from where she is believed to have been abducted. She was found off an unnamed dirt road which was used for access to the woods by hunters and fishermen in the Brimfield area. The longitude and latitude for Brimfield, MA are 72W12 and 42N07. Investigators have ruled her death as a homicide, but there are no suspects.

On May 13, 1993, more than 500 people grieved on the overcast day when Piirainen's small white casket was carried into the church. Only six months before she had made her first confession in the same church.

Illustration 12 - Holly Piirainen Missing Person Flyer (Piirainen)

Piirainen's Natal Chart

Piirainen's natal chart shows a grand trine between the North Node at 02:52 Cancer R in the third house of everyday environment, Saturn at 03:57 degrees of Scorpio in the seventh house of one-to-one relationships, and Mars at 01:39 Pisces in the eleventh house of friends. Her Ascendant at 22:46 Aries also squares Mercury at 21:00 Capricorn R in the tenth house of public recognition. Jupiter is placed at the critical degree of 04:36 Sagittarius in the eighth house of injury, death and regeneration.

The Sun at 29:00 Capricorn in the tenth house of public recognition sextiles Piirainen's natal Moon at 28:46 Pisces in the twelfth house of secret enemies. The Sun also squares Pluto at 29:29 Libra in the seventh house of one-to-one relationships.

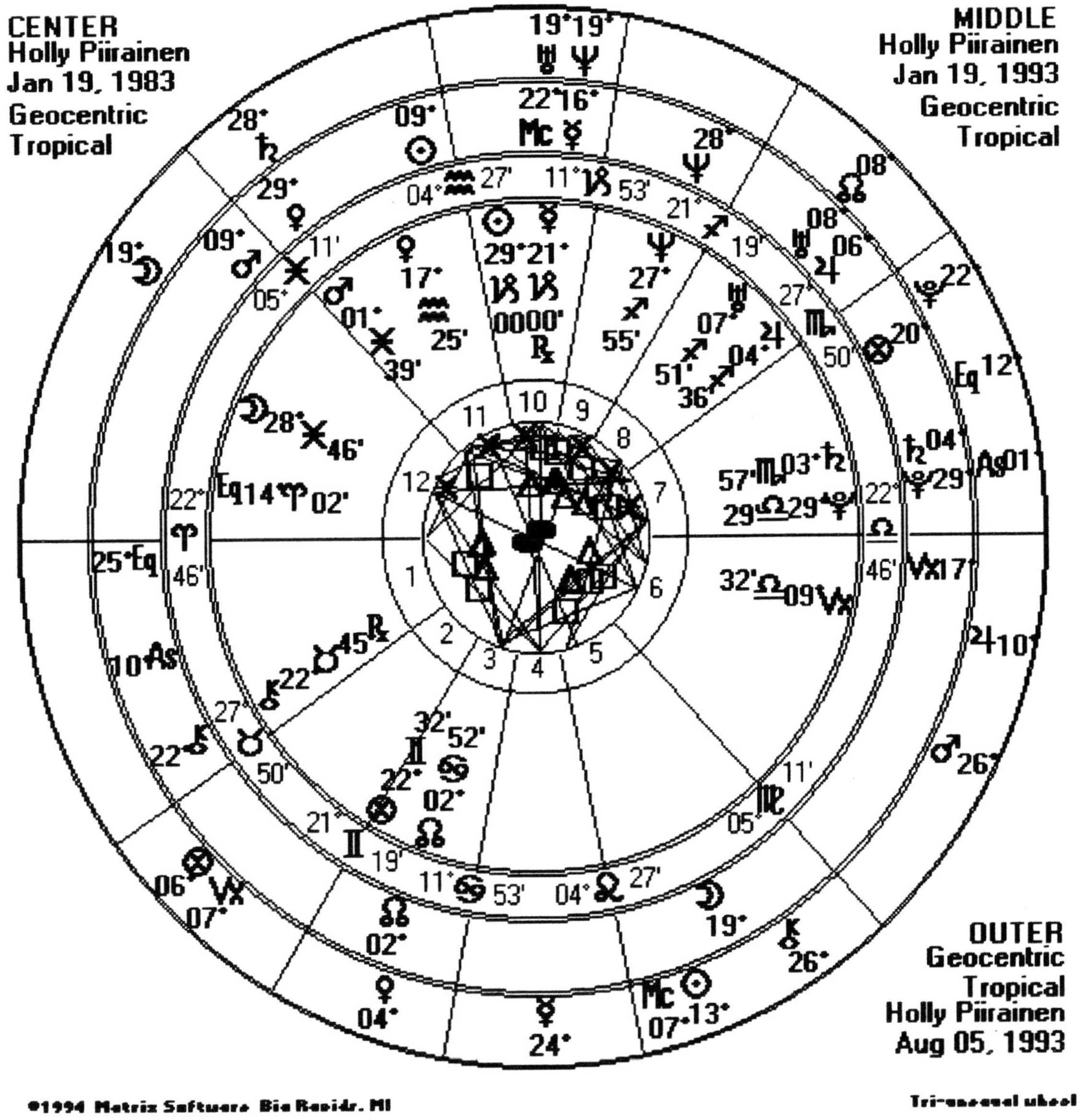

Chart 14 - Holly Piirainen's Three-ring Chart with Natal, Progressed Chart for 1993 and Transits for August 5, 1993

The Progressed Chart for 1993

Piirainen's progressed Sun at the critical degree of 09:10 Aquarius in the natal eleventh house of friends and acquaintances sextiles progressed Uranus at 08:16 Sagittarius in the natal eighth house of injury and death. The progressed Sun also squares the progressed Ascendant at 10:13 Taurus.

Progressed Venus at 29:55 Aquarius in the natal eleventh house of friends and acquaintances trines natal Pluto at 29:29 Libra and progressed Pluto at 29:32 Libra both in the natal seventh house of one-to-one relationships. Progressed Neptune at 28:14 Sagittarius in the natal ninth house also sextiles natal and progressed Pluto.

Piirainen's Transits Interpreted for August 5, 1993

There are two transiting aspects that should be noted in Piirainen's charts. Transiting Pluto at 22:43 Scorpio in the natal seventh house of one-to-one relationships forms an applying inconjunct to the natal Ascendant at 22:46 Aries. Transiting Saturn at 28:01 Aquarius R in the eleventh house of friends and acquaintances is separating from a sextile with natal Neptune at 27:55 Sagittarius in the ninth house of the unconscious mind.

Frances Sakoian and Louis Acker authors of *Predictive Astrology*, interpret the Saturn sextile Neptune aspect as insight into the subconscious mind. They suggest that one is able to tap resources of the memory, which might even go back to previous embodiments. They further interpret that the individual is provided with profound insights that are useful to him or her in the present.

A Partial Solar Eclipse occurred on May 21, 1993 at 00:32 Gemini. The eclipse falls in Piirainen's second house and can indicate changes in personal resources.

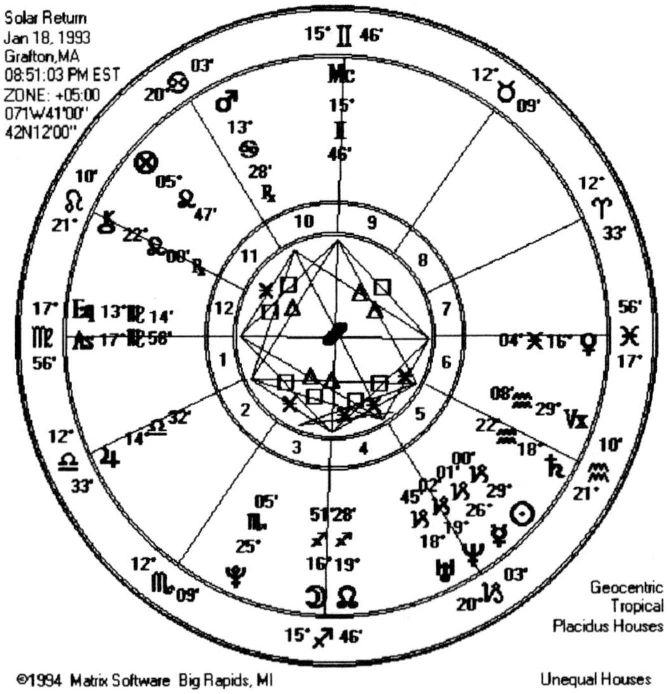

Chart 15 - Holly Piirainen's Solar Return Chart for 1993

The Solar Return Chart for 1993 (Grafton, MA)

Piirainen's solar return chart for 1993 has three critical degrees. Solar Mercury at 26:01 Capricorn is three degrees away from an exact conjunction with Piirainen's solar

Sun at 29:00 Capricorn both in the fifth house of circumstances affecting romance and creativity. Solar Mars at the critical degree of 13:28 Cancer R in the tenth house of public recognition squares solar Jupiter at 14:32 Libra in the second house of values and resources.

The solar Ascendant at the critical degree of 17:56 Virgo has several aspects. It squares the Moon at 16:51 Sagittarius and the North Node at 19:28 Sagittarius R both in the fourth house of emotional and physical home. The Ascendant trines Uranus at 18:45 Capricorn and Neptune at 19:02 Capricorn both also in the fourth house. It further inconjuncts Saturn at 18:22 Aquarius in the fifth house of circumstances affecting romance and creativity, and opposes Venus at 16:04 Pisces in the sixth house of health and service to others.

It should be noted that both the solar Sun and the Vertex are at 29:00. The Sun is at 29:00 Capricorn in the fifth house and the Vertex is at 29:08 Aquarius in the sixth house.

STATE

TIMES UNION
Albany, New York
Monday, October 25, 1993
B-2

Massachusetts case studied for clues on missing N.Y. girl

FRANKFORT — The discovery of a child's body in the Massachusetts woods may provide insight into the case of a central New York girl who has been missing for two months, State Police said Sunday.

HOLLY PIIRAINEN, 10, of Grafton, Mass., was found slain in a wooded area in that state. Troopers in New York are studying the case for clues to the disappearance of 12-year-old Sara Anne Wood of Frankfort.

Illustration 13 - Albany Times Union October 25, 1993 (Piirainen)

In Conclusion

Piirainen's abduction and murder case is still open. It was thought by authorities that a possible suspect was Lewis Lent, Jr. of North Adams, MA. Lent is also a suspect in the abduction/murder of Sara Ann Wood, 12, of Frankfort, New York, who vanished from the area near her home on August 18, 1993 and is still missing. Lent has since been charged with and has pleaded guilty to the abduction and murder of Jimmy Bernardo, 12, of Pittsfield, MA in 1990. Lewis Lent, Jr. was born on May 21, 1950 in Reynoldsville, NY. The longitude and latitude for Reynoldsville, NY are 076W46 and 42N28.

Birth Data

The birth data is being included for Piirainen's parents for your information and use.

Birth Data for Holly Piirainen:

Birth date:	January 19, 1983		
Birth time:	10:45 AM/EST		
Birth place:	Worcester, MA		
Longitude:	071W48	Latitude:	42N16
Source:	Birth certificate		

Birth Data for Piirainen's Parents:

Mother:	Christina Marie Harrington Piirainen		
Birth date:	August 08, 1962		
Birth time:	09:11 PM/EDT		
Birth place:	Boston, MA		
Longitude:	071W04	Latitude:	42N21
Source:	Lois Rodden's *Data News* #46 April 1994		

Father:	Richard Piirainen		
Birth date:	February 12, 1960		
Birth time:	11:37 AM/EST		
Birth place:	Worcester, MA		
Longitude:	071W48	Latitude:	42N16
Source:	Lois Rodden's *Data News* #46 April 1994.		

Placements and degrees of Piirainen's charts:
Natal Placements

Moon	28:46 Pis	12th	Neptune	27:55 Sag	09th
Sun	29:00 Cap	10th	Pluto	29:29 Lib	07th
Mercury	21.00 Cap R	10th	North Node	02:52 Can R	03rd
Venus	17:25 Aqu	11th	Vertex	09:32 Lib	06th
Mars	01:39 Pis	11th	East Point	14:02 Ari	12th
Jupiter	04:36 Sag	08th	Fortuna	22:32 Can	03rd
Saturn	03:57 Sco	07th	Midheaven	11:53 Cap	
Uranus	07:51 Sag	08th	Ascendant	22:46 Ari	

Eleventh house position is 04:27 Aqu
Twelfth house position is 05:11 Pis

Second house position is 27:50 Tau
Third house position is 21:19 Gem

Critical Degrees: Jupiter

To be noted: The Moon, Sun, and Pluto at or approaching 29 degrees
Intercepted Signs: None

Progressed Placements for 1993

Moon	19:26 Leo	05th	Neptune	28:14 Sag	09th
Sun	09:10 Aqu	11th	Pluto	29:32 Lib	07th
Mercury	16:45 Cap	10th	North Node	02:21 Can R	03rd
Venus	29:55 Aqu	11th	Vertex	17:33 Lib	06th
Mars	09:30 Pis	12th	East Point	25:42 Ari	01st
Jupiter	06:13 Sag	08th	Fortuna	20:29 Sco	07th
Saturn	04:16 Sco	07th	Midheaven	22:03 Cap	10th
Uranus	08:16 Sag	08th	Ascendant	10:13 Tau	01st

Critical Degrees: Sun
To be noted: Venus and Pluto at 29 degrees of Libra

Transiting Placements for August 5, 1993 (12:30 PM/EDT)

Moon	19:00 Pis	12th	Uranus	19:17 Cap R	10th
Sun	13:15 Leo	05th	Neptune	19:08 Cap R	10th
Mercury	24:05 Can	04th	Pluto	22:43 Sco	07th
Venus	04:18 Can	03rd	North Node	08:57 Sag R	08th
Mars	26:01 Vir	06th	Vertex	07:30 Gem	02nd
Jupiter	10:29 Lib	06th	East Point	12:23 Sco	07th
Saturn	28:01 Aqu R	11th	Fortuna	06:50 Gem	02nd

Critical Degrees: None
To be noted: None

Solar Return Placements for 1993 (Grafton, MA)

Moon	16:51 Sag	04th	Neptune	19:02 Cap	04th
Sun	29:00 Cap	05th	Pluto	25:05 Sco	03rd
Mercury	26:01 Cap	05th	North Node	19:28 Sag R	04th
Venus	16:04 Pis	06th	Vertex	29:08 Aqu	06th
Mars	13:28 Can R	10th	East Point	13:14 Vir	12th
Jupiter	14:32 Lib	02nd	Fortuna	05:47 Leo	11th
Saturn	18:22 Aqu	05th	Midheaven	15:46 Gem	
Uranus	18:45 Cap	04th	Ascendant	17:56 Vir	

Eleventh house position is 20:03 Can
Twelfth house position is 21:10 Leo
Second house position is 12:33 Lib
Third house position is 12:09 Sco

Critical Degrees: Mercury in the fifth house of creativity and circumstances affected by romance, Mars in the tenth house of public recognition and the Ascendant
To be noted: Sun and the Vertex are at 29 degrees
Intercepted Signs: None

Polly Hannah Klaas
1981 - 1993

Twelve-year-old Polly Klaas was having a slumber party with two of her girlfriends on the night of October 1, 1993. The girls were in Klaas's bedroom in her Petaluma, CA home. She was abducted by an intruder at approximately 10:40 PM/PDT. The longitude and latitude for Petaluma, CA are 122W39 and 38N14. About twenty or thirty minutes later, Klaas's friends, who had been threatened by the intruder felt safe enough to waken her mother. The kidnapping was reported to the Petaluma police at 11:03 p.m.

At 11:42 p.m., a woman reported a trespasser on her ranch just east of Santa Rosa. Deputies stopped to help Richard Allen Davis whose white car was stuck in a drainage ditch. Allen's manner puzzled them, but a check for outstanding warrants and a search of his car turned up nothing. They were also not alerted to the recent abduction of Klaas, and therefore, they helped drag Davis's car free.

Finding of girl's body breaks searchers' hearts

Man arrested in kidnapping, murder of Calif. girl arraigned

Police blamed for mix-up

Kidnapping suspect could have been arrested earlier

2 schoolmates detail Polly Klaas' abducti

MONDAY, DECEMBER 6, 1993

Tears for Polly Klaas

Girl's death leaves town asking why

Illustration 14 - Albany Times Union (Klaas)

Hours after the kidnapping, hundreds of volunteers began to work together in a search effort for Klaas. This effort eventually stretched across the country. Included in this effort was actress Winona Ryder, whose hometown is Petaluma. She offered a $200,000 reward for information leading to the safe return of the young girl. More than eight million missing person flyers were circulated around the world with Klaas's picture and a sketch of her abductor on them. On October 14, 1993, CBS's *Eye to Eye* with Connie Chung featured the story of Klaas's abduction.

More than two months went by without a trace of Klaas emerging. The deputies had not yet made the connection to their meeting with Davis. On October 19, Davis was booked on a drunken driving charge about 70 miles north of Petaluma. He was released within 24 hours. On November 15, Davis failed to report to the parole office in San Mateo County, south of San Francisco.

On November 28, the same woman who had reported the man trespassing on her ranch called the deputies again to say she had found a sweat shirt and other items. It was at this point, that police made the connection with the trespasser and pulling the car from the ditch. The trespasser was identified to be Davis and the clothing belonged to Klaas. Davis, 39, then became the prime suspect and in addition a match of his palm print was made with the one found in Klaas's bedroom. Davis was arrested on November 30, 1993 and charged with Klaas's kidnapping and murder. Davis was a twice-convicted kidnapper who had been paroled in June after serving eight years in prison.

Illustration 15 - Polly Klaas's Missing Person Flyer (Klaas)

Klaas's body was found late on Saturday, December 4, 1993. Although the people of Petaluma mourned the loss of this young girl, they were comforted by the knowledge that they did all they could to aid in her search. Since Klaas's abduction, her community has educated itself about this type of crime and learned some basic rules in the hopes of preventing another such kidnapping.

Klaas's Natal Chart

In Klaas's natal chart, the aspects with the closest orbs is her natal Sun at the critical degree of 13:19 Capricorn in the tenth house suggesting that she would gain public recognition. It inconjuncts her natal North Node at 12:22 Leo R in the sixth house of health. Natal Jupiter at 09:42 Libra conjuncts natal Saturn at 09:35 Libra in the seventh house of others and public enemies, and natal Neptune at 23:09 Sagittarius in the tenth house sextiles natal Pluto at 24:11 Libra in the seventh house.

Klaas's natal house cusps should also be noted. The second house cusp is 29:01 Aries, the third house is approaching 29 degrees at 28:47 Taurus, the fifth house is at the critical degree of 13:38 Cancer and sixth house is approaching the critical degree of 09:00 Leo at 08:58 Leo.

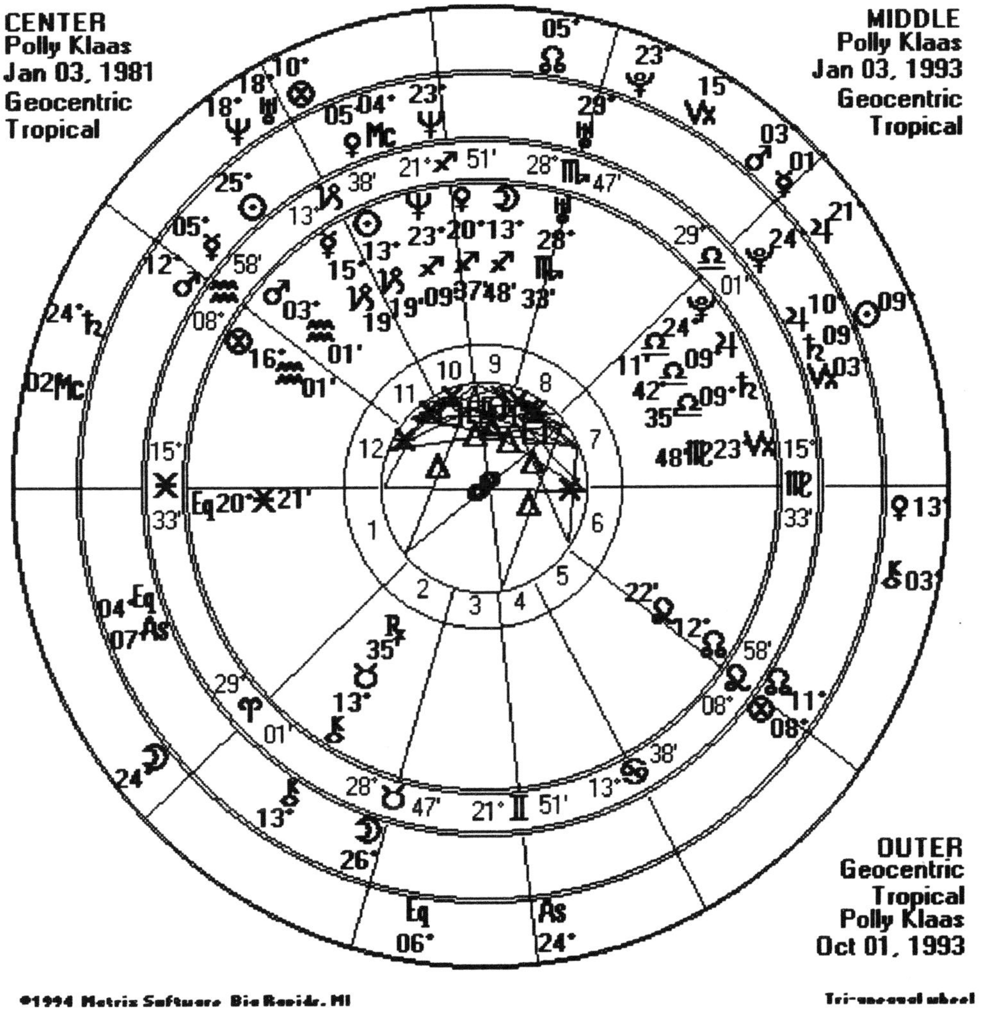

Chart 16 - Polly Klaas's Three-ring Chart with Natal, Progressed Chart for 1993 and Transits for October 01, 1993.

The Progressed Chart for 1993

Progressed Uranus at 29:05 Scorpio in the natal ninth house of abstract mind and spiritual transformation forms two aspects. The first is an opposition to the progressed Moon at 26:41 Taurus in the natal second house of values and finances, and secondly is a sextile to the progressed Sun at 25:33 Capricorn in the natal eleventh house of hopes, wishes and friends.

The progressed Sun at 25:33 Capricorn in the eleventh house trines the progressed Moon at 26:41 Taurus in the natal second house. The progressed Moon inconjuncts Pluto at 24:19 Libra in the natal seventh house of others and public enemies. A T-square between progressed Venus at 05:37 Capricorn in the natal tenth house of public recognition conjuncts the progressed Midheaven at 04:05 Capricorn and squares the progressed Ascendant at 07:17 Aries. The Ascendant opposes progressed Saturn at 09:47 Libra in the natal seventh house.

Progressed Mars at 12:28 Aquarius in the natal twelfth house of unknown enemies opposes the North Node at 11:44 Leo R in the natal fifth house of romance. The progressed Neptune at 23:34 Sagittarius in the natal tenth house of public recognition sextiles progressed Pluto at 24:19 Libra in the natal seventh house of others and public enemies.

Klaas's Transits for October 1, 1993

Transiting Saturn at 24:12 Aquarius R in the natal twelfth house of unknown enemies retrogrades back to trine natal Pluto at 24:11 Libra in the seventh house of others and public enemies. Transiting Jupiter at 21:33 Libra in the seventh house forms an exact sextile with the natal Midheaven at 21:51 Sagittarius on October 03, 1993.

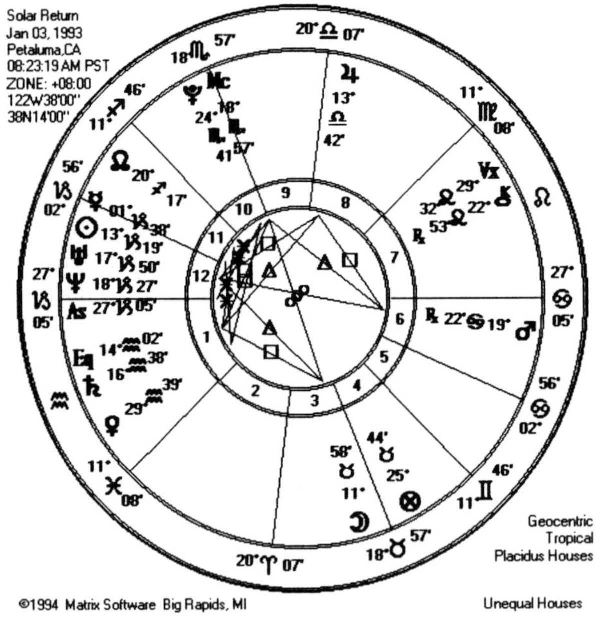

Chart 17 - Polly Klaas's Solar Return Chart for 1993

The Partial Solar Eclipse on May 21, 1993 occurred at 00:32 Gemini and falls in Klaas's third house which rules her everyday environment. This can indicate changes in relationships with neighbors and separation from a close relative.

The Solar Return Chart for 1993 (Petaluma, CA)

Klaas's solar return chart for 1993 has a yod comprised of solar Saturn at 16:38 Aquarius in the first house of self, solar Mars at 19:22 Cancer R in the sixth house of health, and the solar North Node at 20:17 Sagittarius R in the eleventh house of friends and acquaintances. Aquarius is intercepted in the first house of self and Libra in the seventh house of others. Mars also trines the solar Midheaven at 18:57 Scorpio and opposes both Uranus at 17:50 Capricorn and Neptune at 18:27 Capricorn in the twelfth house of secret enemies.

The solar Sun at the critical degree of 13:19 Capricorn in the twelfth house forms a square with solar Jupiter at the critical degree of 13:42 Libra in the eight house of death and sex. Solar Venus at 29:39 Aquarius in the first house of self opposes the solar Vertex at 29:32 Leo in the seventh house of others.

Davis's Natal Chart

Davis's Cancer stellium in the eighth house of death, regeneration and sex is comprised of natal Jupiter, Mercury, Vertex, the Moon, Venus, South Node and Uranus. Opposing the stellium from the second house of values are natal Mars and the North Node. Mercury at 04:46 Cancer and Jupiter at 02:09 Cancer both in the eighth house of other's resources trines Saturn at 03:32 Scorpio R in the karmic twelfth house.

Virgo is intercepted in the tenth house of public recognition and Pisces is intercepted in the fourth house of home and personal matters.

The Progressed Chart for 1993

In Davis's progressed chart there is also a Cancer stellium in the natal seventh house of one-to-one relationships which is comprised of progressed Mercury, Jupiter, the South Node and Uranus in the natal eighth house of death, regeneration and sex. Opposing the stellium from the natal first house of self are the natal and progressed North Node.

Progressed Venus at 28:30 Leo in the natal ninth house of the law and abstract mind trines progressed Mars 27:41 Sagittarius R in the natal second house of finances and values. Progressed Saturn at 02:40 Scorpio, the sign ruling sex, is in the twelfth house of self-undoing and opposes progressed Fortuna or Misfortuna at 02.28 Taurus in the sixth house of health and service.

Davis's Transits Interpreted for October 1, 1993

Transiting Uranus at 18:14 Capricorn forms a close conjunction with transiting Neptune at 18:23 Capricorn in the natal second house of values.

The Partial Solar Eclipse on May 21, 1993 occurred at 00:32 Gemini and falls in Davis's seventh house of one-to-one relationships indicating changes in this area.

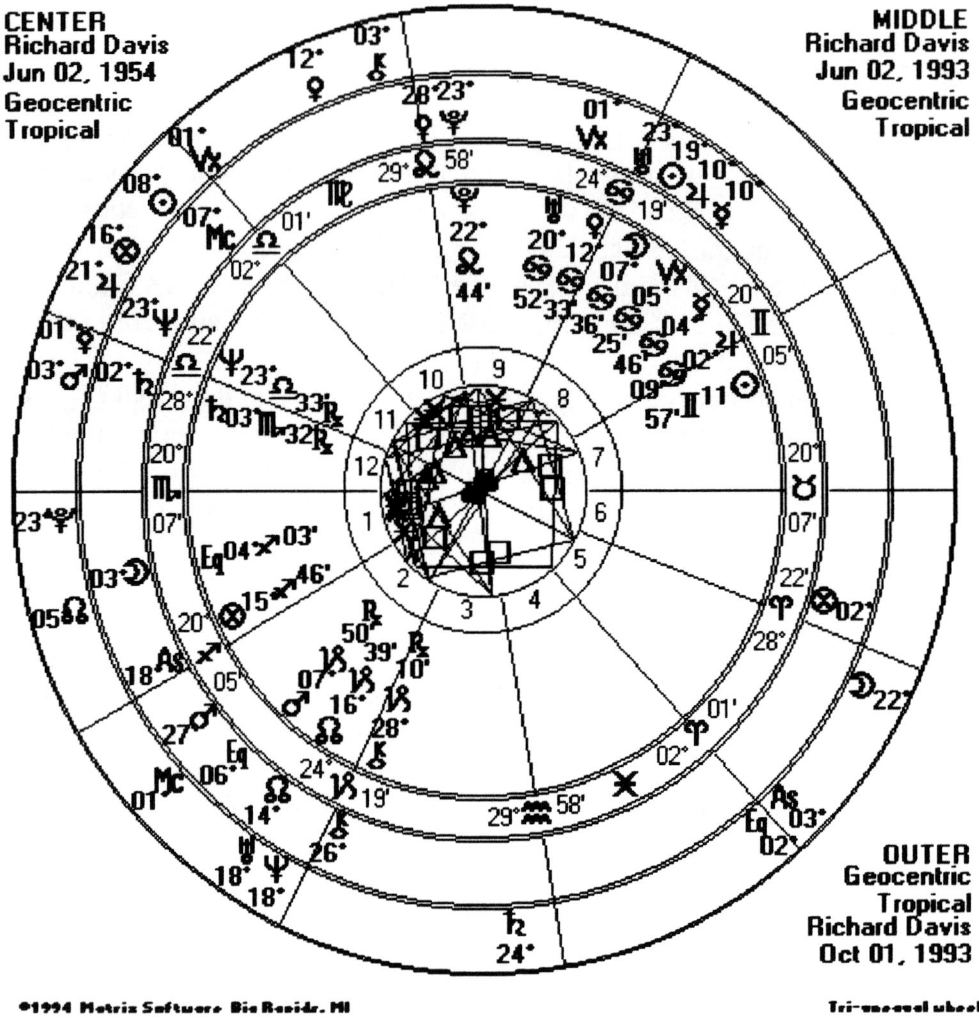

Chart 18 - Richard Allen Davis's Three-ring Chart with Natal, Progressed Chart for 1993 and Transits for October 1, 1993

The Solar Return Chart for 1993 (Cloverdale, CA)

Solar Mercury is at the critical degree of 00:34 Cancer in the second house of values. Mercury trines Saturn at 00:17 Pisces in the eleventh house of friends and inconjuncts the Midheaven at 02:11 Aquarius. Solar Venus is at the critical degree of 26:25 Aries in the twelfth house of self-undoing and imprisonment. The aspect in closest orb to Venus is its opposition to Fortuna and Misfortuna at 25:30 Libra in the sixth house of health.

Virgo, ruling the young maiden, is intercepted in Davis's solar return chart in the fifth house of love pursuits and Pisces is intercepted in the eleventh house of friends and acquaintances. The 29th degree of Leo is also found on the fifth house cusp of children and love relationships and 29 degrees Aquarius on the eleventh house cusp of friends and acquaintances. A need to focus on both love relationships and friendships is indicated.

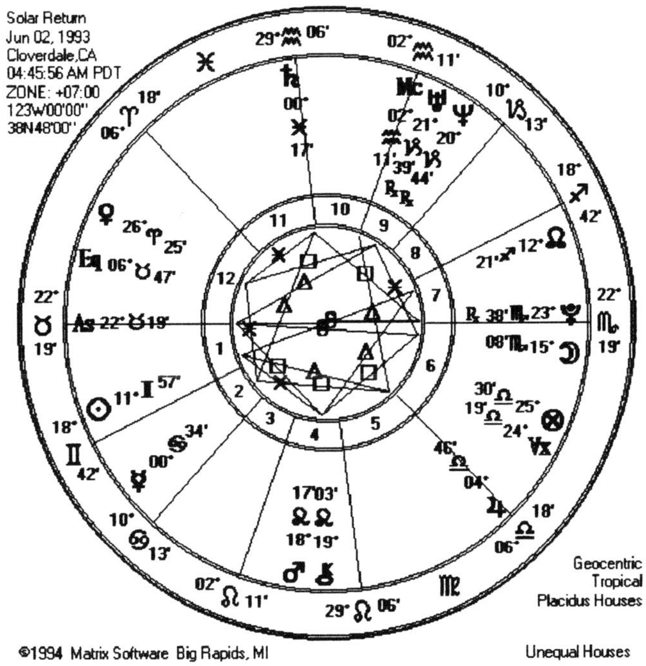

Chart 19 - Richard Davis's Solar Chart for 1993

In Conclusion

On June 18, 1996, a six-man, six-woman jury took 21 hours to decide that Davis was guilty of Klaas's death. The jury then had two choices as possible penalties: death or life without parole. The jury ignored the defense attorney's pleas to show mercy for Davis's troubled childhood and decided on August 5, 1996 that he should be executed.

Birth Data for Polly Klaas:

Birth date:	January 03, 1981		
Birth time:	01:41 AM/PST		
Birth place:	San Francisco, CA		
Longitude:	122W25	Latitude:	37N46
Source:	Birth certificate		

Birth Data for Richard Allen Davis:

Birth date:	June 02, 1954		
Birth time:	06:34 PM/PDT		
Birth place:	San Francisco, CA		
Longitude:	122W25	Latitude:	37N46
Source:	Lois Rodden's *Data News* #45 February 1994		

Placements and degrees of Klaas's charts:
Natal Placements

Moon	13:48 Sag	09th	Neptune	23:09 Sag	10th
Sun	13:19 Cap	10th	Pluto	24.11 Lib	07th
Mercury	15:19 Cap	11th	North Node	12:22 Leo R	06th
Venus	20:37 Sag	09th	Vertex	23:48 Sco	07th
Mars	03:01 Aqu	11th	East Point	20:21 Pis	01st
Jupiter	09:42 Lib	07th	Fortuna	16:01 Aqu	12th
Saturn	09:35 Lib	07th	Midheaven	21:51 Sag	
Uranus	28:33 Sco	08th	Ascendant	15:33 Pis	

Eleventh house position is 13:38 Cap
Twelfth house position is 08:58 Aqu
Second house position is 29:01 Ari
Third house position is 28:47 Tau

Critical Degrees: Sun
To be noted: Klaas's house cusps are second house 29:01 Aries, third house 28:42 Taurus which is approaching 29 degrees, fifth house 13:38 Cancer and sixth house 08:58 Leo approaching 09 degrees
Intercepted Signs: None

Progressed Placements for 1993

Moon	26:41 Tau	02nd	Neptune	23:34 Sag	10th
Sun	25:33 Cap	11th	Pluto	24:19 Lib	07th
Mercury	05:15 Aqu	11th	North Node	11:44 Leo R	06th
Venus	05:37 Sag	09th	Vertex	03:06 Lib	07th
Mars	12:28 Aqu	12th	East Point	04:51 Ari	01st
Jupiter	10:16 Lib	07th	Fortuna	08:25 Leo	05th
Saturn	09:47 Lib	07th	Midheaven	04:05 Cap	10th
Uranus	29:05 Sco	09th	Ascendant	07:17 Ari	01st

Critical Degrees: None
To be noted: Uranus at 29 degrees in the ninth house of long journeys

Transiting Placements for October 1, 1993 (10:45 PM/PST)

Moon	24:49 Ari	01st	Uranus	18:14 Cap	11th
Sun	09:03 Lib	07th	Neptune	18:23 Cap	11th
Mercury	01:33 Sco	08th	Pluto	23:42 Sco R	08th
Venus	13:01 Vir	06th	North Node	05:54 Sag	09th
Mars	03:29 Sco	08th	Vertex	15:01 Sco	08th
Jupiter	21:33 Lib	07th	East Point	06:39 Gem	03rd
Saturn	24:12 Aqu R	12th	Fortuna	10:34 Cap	10th

Critical Degrees: None
To be noted: None

Solar Return Placements for 1993 (Petaluma, CA)

Moon	11:56 Tau	03rd	Neptune	18:27 Cap	12th
Sun	13:19 Cap	12th	Pluto	24:41 Sco	10th
Mercury	01:38 Cap	11th	North Node	20:17 Sag R	11th
Venus	29:39 Aqu	01st	Vertex	29:32 Leo	07th
Mars	19:22 Can	06th	East Point	14:02 Aqu	01st
Jupiter	13:42 Lib	08th	Fortuna	25:44 Leo	07th
Saturn	16:38 Aqu	01st	Midheaven	18:57 Sco	
Uranus	17:50 Cap	12th	Ascendant	27:07 Cap	

Eleventh house position is 11:46 Sag
Twelfth house position is 02:56 Cap
Second house position is 11:08 Pis
Third house position is 20:07 Ari

Critical Degrees: Sun and Jupiter
To be noted: Venus at 29 degrees in the first house and Vertex at 29 degrees in the seventh house
Intercepted Signs: Leo in the seventh house of one-to-one relationships and Aquarius in the first house of self

Placements and degrees of Richard Allen Davis's charts:
Natal Placements

Moon	07:36 Can	08th	Neptune	23:33 Lib R	11th
Sun	11:57 Gem	07th	Pluto	22:44 Leo	09th
Mercury	04:46 Can	08th	North Node	16:39 Cap R	02nd
Venus	12:33 Can	08th	Vertex	05:25 Can	08th
Mars	07:50 Cap R	02nd	East Point	04:03 Sag	01st
Jupiter	02:09 Can	08th	Fortuna	15:46 Sag	01st
Saturn	03:32 Sco R	12th	Midheaven	29:58 Leo	
Uranus	20:52 Can	08th	Ascendant	20:07 Sco	

Eleventh house position is 02:01 Lib
Twelfth house position is 28:22 Lib
Second house position is 20:05 Sag
Third house position is 24:19 Cap

Critical Degrees: None
To be noted: Cancer stellium in the eighth house, the Midheaven at 29 degrees
Intercepted Signs: Virgo in the tenth house and Pisces in the fourth house

Progressed Placements for 1993

Moon	03:10 Sag	01st	Neptune	23:18 Lib	11th

Sun	19:11 Can	08th	Pluto	23:33 Leo	09th
Mercury	10:25 Can R	08th	North Node	14:35 Cap R	02nd
Venus	28:30 Leo	09th	Vertex	01:56 Leo	09th
Mars	27:41 Sag R	02nd	East Point	06:04 Cap	02nd
Jupiter	10:57 Can	08th	Fortuna	02:28 Can	06th
Saturn	02:40 Sco	12th	Midheaven	07:12 Lib	11th
Uranus	23:05 Can	08th	Ascendant	18:28 Sag	01st

Critical Degrees: None
To be noted: None

Transiting Placements for October 1, 1993

Moon	22:45 Ari	05th	Uranus	18:14 Cap	02nd
Sun	08:53 Lib	11th	Neptune	18:23 Cap	02nd
Mercury	01:19 Sco	12th	Pluto	23:42 Sco	01st
Venus	12:48 Vir	10th	North Node	05:54 Sag R	01st
Mars	03:22 Sco	12th	Vertex	01:18 Lib	11th
Jupiter	21:31 Lib	11th	East Point	02:02 Ari	05th
Saturn	24:12 Aqu R	03rd	Fortuna	16:57 Lib	11th

Critical Degrees: None
To be noted: None

Solar Return Placements for 1993 (Cloverdale, CA)

Moon	15:08 Sco	06th	Neptune	20:44 Cap R	09th
Sun	11:57 Gem	01st	Pluto	23:38 Sco R	07th
Mercury	00:34 Can	02nd	North Node	12:21 Sag R	07th
Venus	26:25 Ari	12th	Vertex	24:19 Lib	06th
Mars	18:17 Leo	04th	East Point	06:47 Tau	12th
Jupiter	04:46 Lib	05th	Fortuna	25:30 Lib	06th
Saturn	00:17 Pis	11th	Midheaven	02:11 Aqu	
Uranus	21:39 Cap R	09th	Ascendant	22:19 Tau	

Eleventh house position is 29:06 Aqu
Twelfth house position is 06:18 Ari
Second house position is 18:42 Gem
Third house position is 10:13 Can

Critical Degrees: Mercury, Venus
To be noted: Saturn at 00 degrees of Pisces, 29 degrees of Leo on the fifth house cusp and 29 degrees of Aquarius on the eleventh house cusp
Intercepted Signs: Virgo in the fifth house and Pisces in the eleventh house

Steven Stayner
1965 - 1989

Many may remember the NBC miniseries in May 1989 entitled "*I Know My First Name Is Steven.*" It told the story of seven-year-old Steven Stayner, who was kidnapped on December 4, 1972 in Merced, California. Steven was forced to live in a Manchester, CA cabin as the sexually abused foster son of his abductor, Kenneth Eugene Parnell. Parnell was a 40-year-old Texas drifter.

Illustration 16 - Steven Stayner's Missing Child Poster (Stayner)

Stayner's sexual abuse continued until he walked into a Ukiah, CA police station on March 1, 1980 carrying a five-year-old boy, Timmy, on his back. Timmy was another victim of Parnell. Stayner said he did not want to see the young boy subjected to the same abuse he had experienced for the past several years. The longitude and latitude for Ukiah, are 123W12 and 39N09.

Parnell was arrested and convicted of two counts of kidnapping. In February 1982, he was sentenced to eight years and eight months in prison. Parnell was paroled on April 5, 1988, and later finished out the terms of his parole in a Berkeley, California rooming house.

On September 16, 1989, Stayner was killed in a hit-and-run accident when a car pulled into the path of his motorcycle on a highway between Merced and Atwater, CA. Stayner, who was riding without a helmet, suffered massive head injuries and was pronounced dead on arrival at Merced Community Medical Center. He was survived by his wife, Jody Edmondson Stayner, whom he married in 1985, a daughter, Ashley, age three, and a son, Steven, Jr., age two.

Steven's Natal Chart

Stayner's natal Moon at 27:53 Scorpio in the ninth house of abstract thought inconjuncts both the Sun at 28:10 Aries and Venus at 29:48 Aries in the second house of finances and values. The Moon also opposes Jupiter at the ominous degree of 29:05 Taurus in the third house of everyday environment.

Natal Saturn at 13:25 in the intercepted sign of Pisces in the first house of self opposes Uranus at 11:05 Virgo R and Pluto R at 14:01 in the intercepted sign of Virgo in the seventh house of one-to-one relationships. Natal Uranus at 11:05 Virgo R conjuncts the Vertex at 11:21 Virgo in the seventh house.

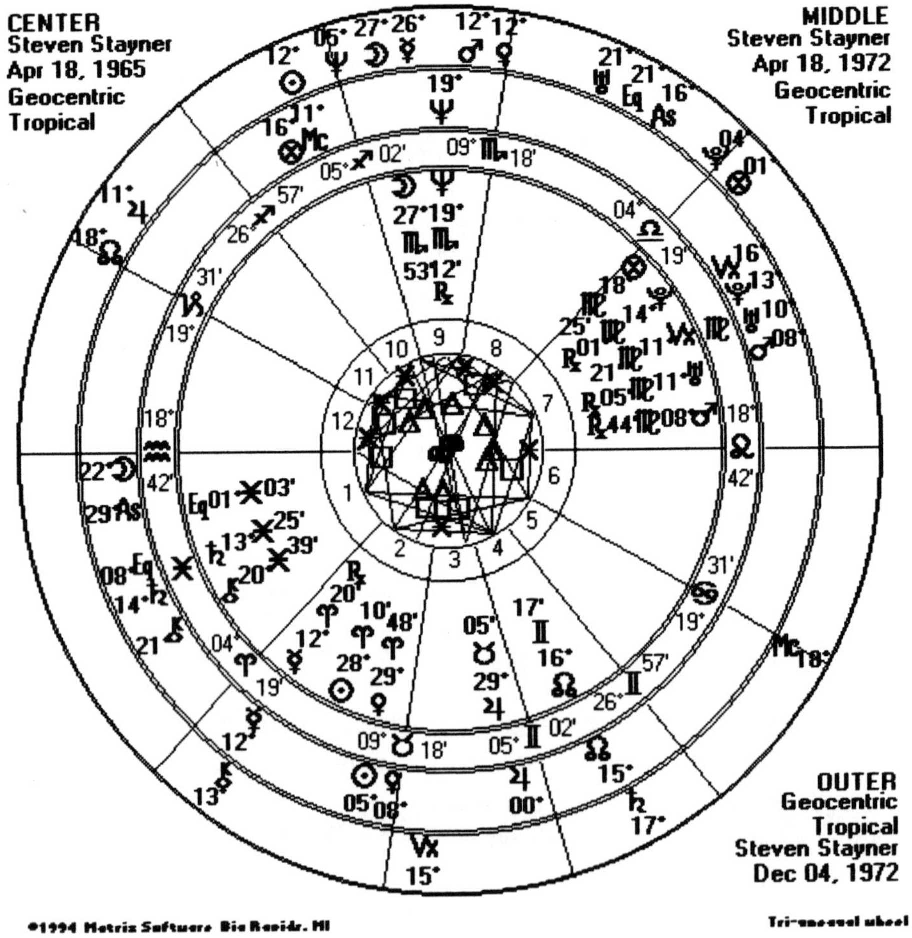

Chart 20 - Stayner's Three-ring Chart with Natal, Progressed Chart for 1972 and Transits for December 4, 1972

The Progressed Chart for 1972

Stayner was abducted on December 4, 1972. In his progressed chart, the Ascendant has progressed to 29:03 Aquarius. Progressed Jupiter at 00:38 Gemini in the natal third house of everyday environment squares his progressed Ascendant. Progressed Mercury at 12:12 Aries in the natal second house of values inconjuncts both progressed Uranus at 10:56 Virgo R and Pluto at 13:54 Virgo R in the natal seventh house of one-to-one relationships.

Progressed Venus at 08:27 Taurus in the second house forms a close trine with progressed Mars at 08:54 Virgo in the natal seventh house. Progressed Saturn at 14:05 Pisces in the natal first house opposes both natal Pluto at 14:01 Virgo R and progressed Pluto at 13:54 Virgo R. This opposition also forms a mutable grand cross with the North Node at 15:55 Gemini R in the natal fourth house and the South Node at 15:55 Sagittarius R in the natal tenth house of authority figures. Mutable planets are adaptable, and they allowed Stayner to adapt to his abduction and the ensuing lifestyle.

Stayner's Transits Interpreted for December 4, 1972

There are three close conjunctions of the transit to transit planets. The transiting Moon at 27:06 Scorpio conjuncts transiting Mercury at 26:23 Scorpio R in the natal ninth house of abstract thought. Transiting Venus at 12:12 Scorpio forms a conjunction with transiting Mars at 12:19 Scorpio in the natal ninth house. Transiting Uranus at 21:46 Libra conjuncts the transiting East Point at 21:45 Libra both in the natal eighth house of shared resources and sex.

Transiting Saturn is at the critical degree of 17:29 Gemini R in the natal fourth house of the emotional and physical home. Transiting Neptune is at 05:15 Sagittarius and crosses the natal Midheaven at 05:02 Sagittarius on November 28, 1972. These placements add secrecy and confusion to Steven's disappearance. Transiting Jupiter at 11:35 Capricorn in the natal eleventh house of friends, hopes and wishes squares natal Mercury at 12:20 Aries R in the natal second house of personal values and finances. Transiting Jupiter in the natal eleventh house of also trines natal Uranus at 11:05 Virgo R in the natal seventh house of one-to-one relationships.

A Total Solar Eclipse on July 10, 1972 occurred at 18:36 Cancer. The eclipse falls in Stayner's fifth house and can indicate changes in romantic ideas, children and creativity.

The Solar Return Chart for 1972 (Merced, CA)

Stayner's solar return chart for 1972 is filled with 29 degree placements. Solar Pluto retrogrades back to 29:59 Scorpio R in the eleventh house of friends, hopes and wishes. The solar Ascendant is at 29:36 Libra, and solar Fortuna or Misfortuna is at 29:02 Sagittarius in the second house of personal finances and values. In addition to the solar Ascendant's placement at 29 degrees, the third house cusp is 29:54 Sagittarius and the ninth house cusp is 29:54 Gemini.

A grand trine is formed by solar Neptune at the critical degree of 04:48 Sagittarius R in the second house of personal values, solar Mercury at 04:21 Aries in the fifth house of romantic relationships and the solar Midheaven at 03:59 Leo in the tenth house of public recognition. Cancer is intercepted in the ninth house of abstract mind and Capricorn is intercepted in the third house of everyday environment.

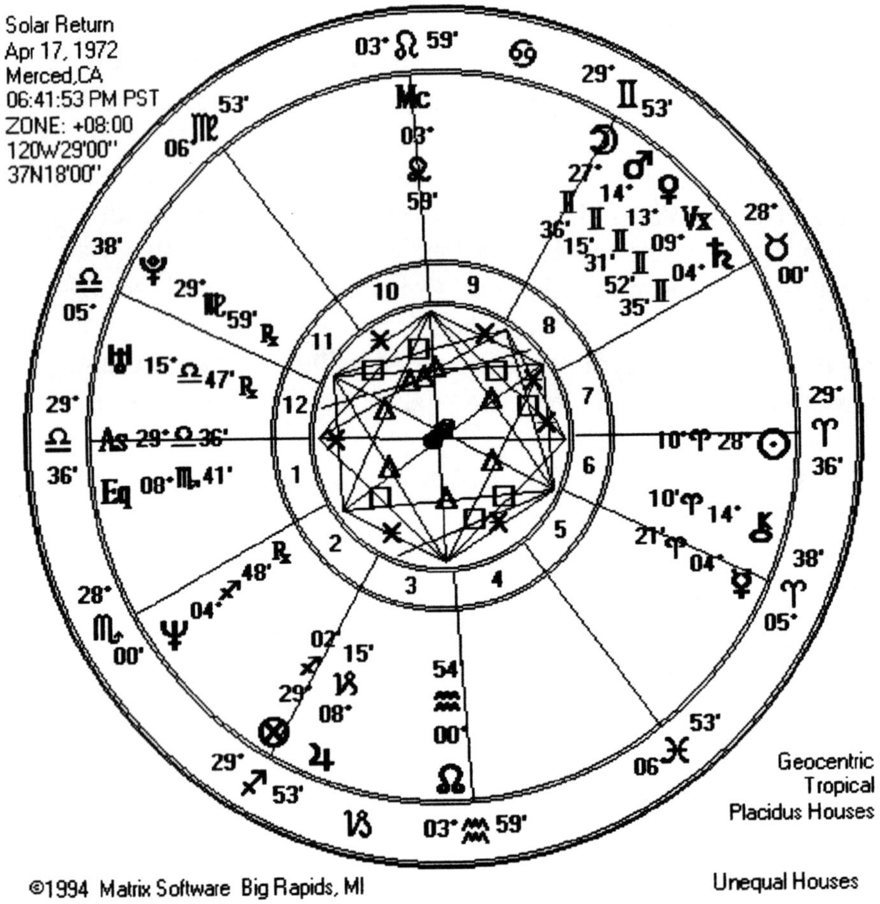

Chart 21 - Stayner's Solar Chart for 1972

Solar Saturn at the critical degree of 04:35 Gemini in the solar eighth house of others and sex opposes Solar Neptune at the critical degree of 04:48 R in the solar second house of values. Solar Saturn also sextiles solar Mercury at 04:21 Aries in the fifth house of romantic relationships.

Progressed Placements for 1979

Stayner walked into a Ukiah, CA police station on March 1, 1980 carrying a five-year-old boy, Timmy, on his back. His progressed chart for 1979 has a mutable grand cross as it did in 1972 when he was abducted. It is formed by progressed Saturn at 14:41 Pisces in the natal first house of self, the progressed North Node at 15:33 Gemini R in the natal fourth house of home and personal affairs, progressed Pluto at 13:49 Virgo R in the natal seven house of one-to-one relationships, and the progressed South Node at 15:33 Sagittarius R in the natal tenth house of authority figures and public recognition.

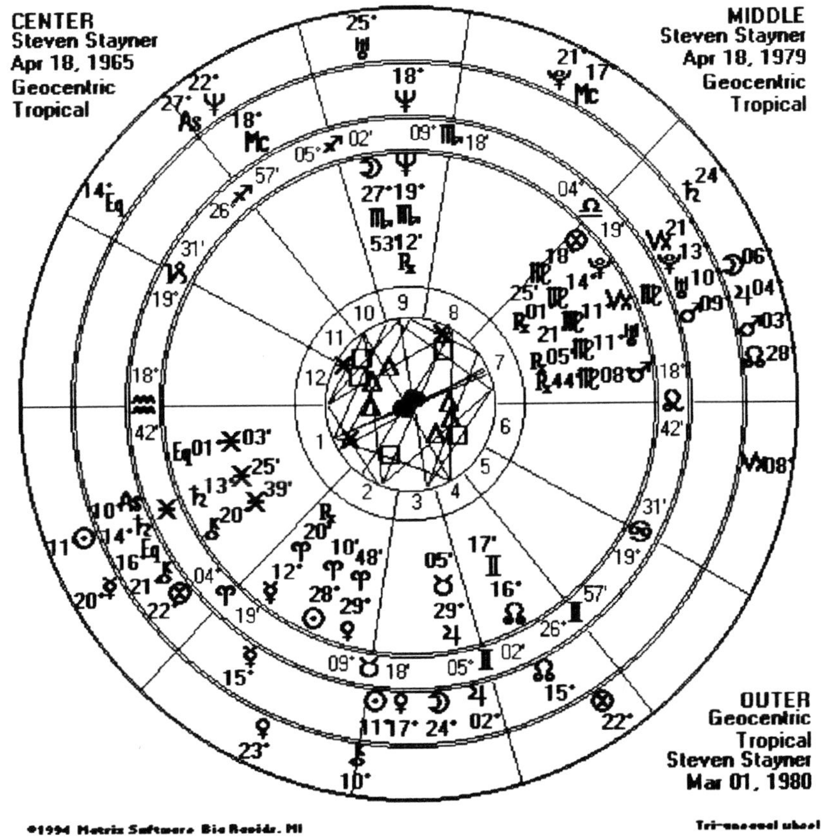

Chart 22 - Stayner's Three-ring Chart with Natal, Progressed Chart for 1979 and Transits for March 1, 1980

Transiting Placements for March 1, 1980

Transiting Moon at 06:15 Virgo, Mars at 03:44 Virgo R and Jupiter at 04:33 Virgo R, all in the seventh house of one-to-one relationships, square the Midheaven at 05:02 Sagittarius. In addition, Jupiter is at the critical degree of 04:33 Virgo. Transiting Mercury at 20:01 Pisces R in the natal first house of self trines natal Neptune at 19:12 Scorpio R in the natal ninth house of abstract thought.

A Total Solar Eclipse on February 16, 1980 occurred at 26:51 Aquarius. The eclipse falls in Stayner's first house and can indicate changes in the ways one expresses his or her identity.

Solar Return Placements for 1979 (Manchester, CA)

The solar Vertex is at the critical degree of 17:16 in the intercepted sign of Sagittarius the solar fifth house of romantic relationships and children. It sextiles solar Pluto at 17:33 Libra R in the solar third house of everyday environment. The solar Moon at 15:45 Capricorn in the solar sixth house of service to others trines the solar North Node at 15:31 Virgo R in the solar second house of personal values.

The solar Uranus at 19:52 Scorpio R in the solar fourth house of personal and domestic matters inconjuncts solar Fortuna or Misfortuna at 19:40 Aries in the solar tenth house of authority figures and public recognition. Jupiter, the planet of expansion and growth, at 29:53 Cancer, ruler of the emotional and physical home, in the karmic twelfth house forms no close aspects except for its conjunction with the solar Ascendant at 02:05 Leo.

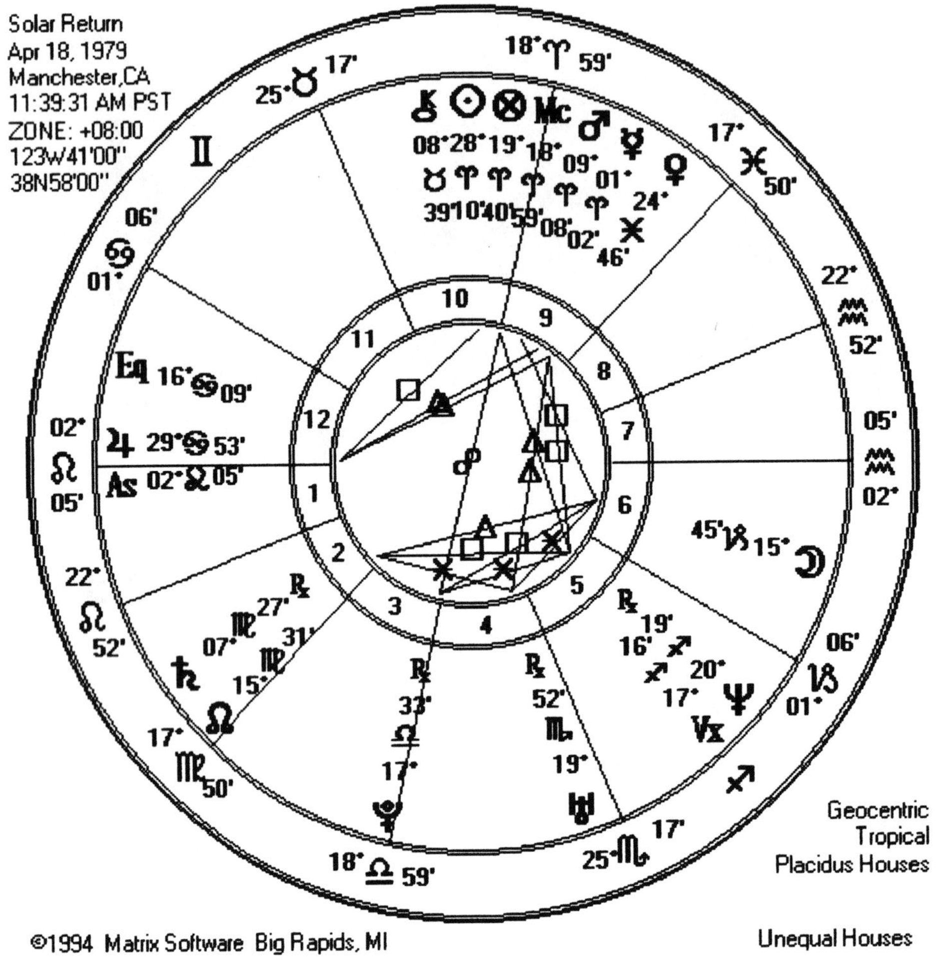

Chart 23 - Stayner's Solar Chart for 1979

Gemini is intercepted in the solar eleventh house of friends, hopes and wishes and Sagittarius is intercepted in the solar fifth house of children, love relationships and creativity.

Progressed Placements for 1989

The progressed Sun is at the critical degree of 21:29 Taurus in the natal third house of everyday environment and travels, and it forms no close aspects. Progressed Mercury at 25:55 Aries is approaching the critical degree of 26 Aries in the natal second house of finances and values. Progressed Jupiter is at the critical degree of 04:30 Gemini in the natal third house, and it forms no close aspects. Caution would be suggested in Stayner's everyday environment and travels, but he was not wearing a helmet on the day of his accident. Aries rules the head and Gemini is the natural ruler of the third house short-distance travels.

A T-square is formed in the progressed chart by progressed Saturn at 15:28 in the intercepted sign Pisces in the natal first house of self. It opposes progressed Pluto at 13:43 of Virgo R which is intercepted in the natal seventh house of one-to-one relationships. These planets square the progressed North Node at 15:01 Gemini R in the natal third house.

Progressed Venus is at the ominous degree of 29:24 Taurus in the natal third house of short-distance travels conjunct natal Jupiter at 29:05 Taurus. Two aspects are formed. One is a sextile to the progressed Ascendant at 27:03 Pisces, and the other is an inconjunct to the progressed Midheaven at 28:20 Sagittarius.

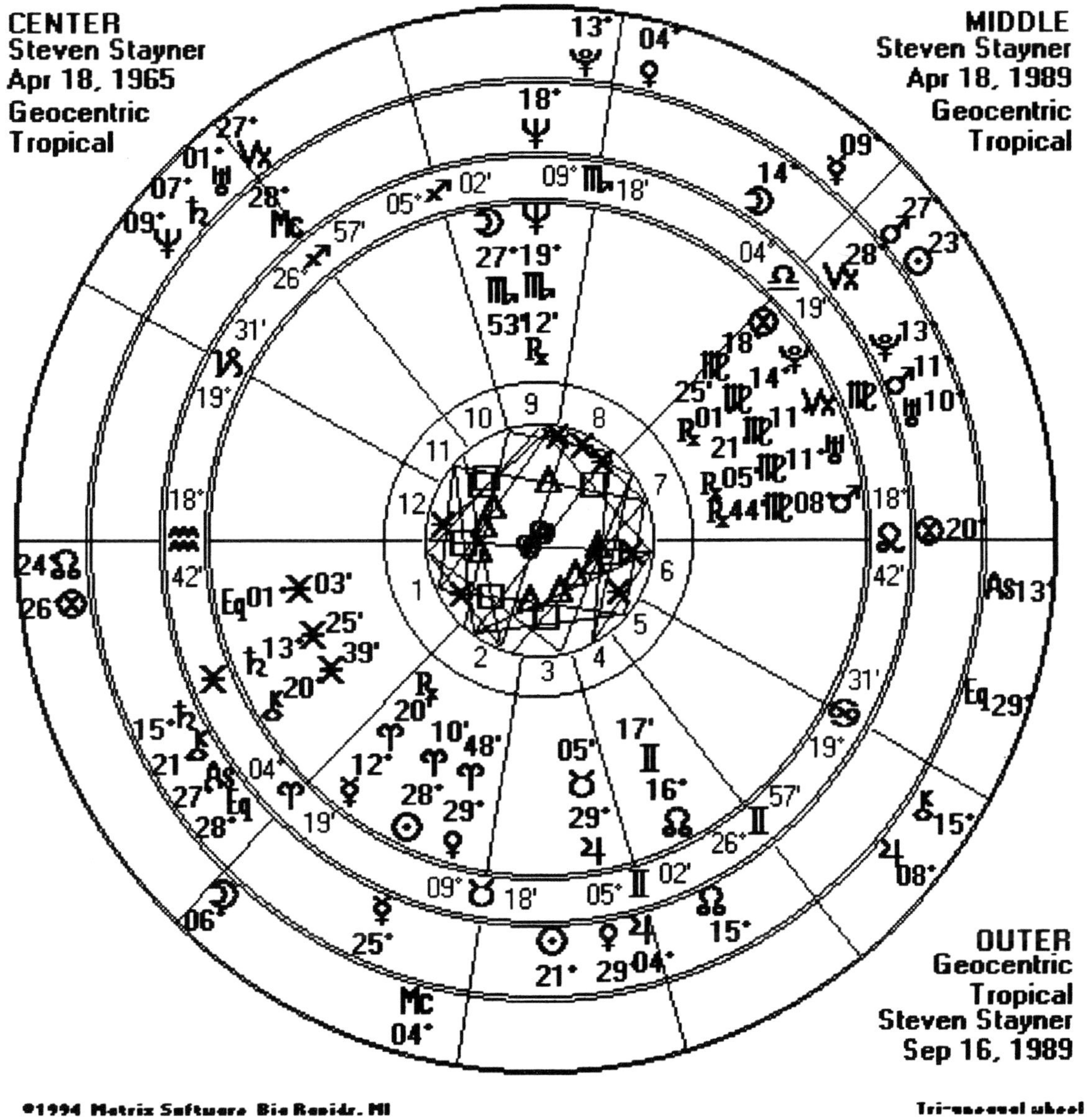

Chart 24 - Stayner's Three-ring Chart with Natal, Progressed Chart for 1989 and Transits for September 16, 1989

Transiting Placements for September 16, 1989

Transiting Pluto at 13:12 Scorpio in the natal ninth house of abstract thought and long-distance travels trines natal Saturn at 13:25 in the first house of self. Transiting Neptune at 09:37 Capricorn R in the natal eleventh house of hopes and wishes trines natal Mars at 08:44 Virgo R in the seventh house of one-to-one relationships. The East Point at 29:56 Cancer in the natal sixth house of health and service squares natal Venus at 29:48 Taurus in the natal second house of personal values.

A Partial Solar Eclipse on August 31, 1989 occurred at 07:48 Virgo. The eclipse falls in the intercepted sign of Virgo in Steven's seventh house and can indicate changes in understanding one-to-one relationships.

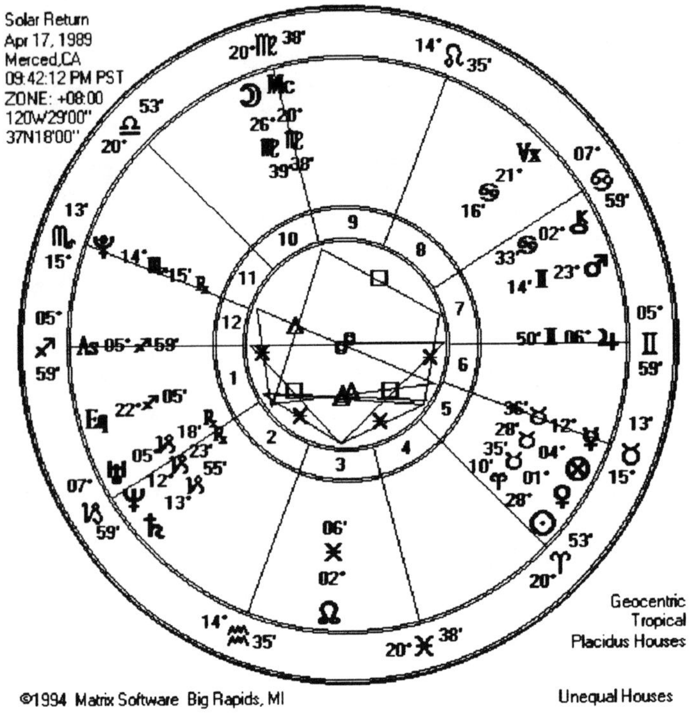

Chart 25 - Stayner's Solar Chart for 1989

Solar Return Placements for 1989 (Merced, CA)

Solar Saturn is at the critical degree of 13:55 Capricorn in the second house of income and personal values. It forms a trine with Mercury at 12:36 Taurus in the fifth house of romantic interests and a sextile to Pluto at 14:15 Scorpio R in the eleventh house of hopes and wishes. It is also in conjunction with solar Neptune at 12:23 Capricorn R in the second house.

In Conclusion

On January 5, 1990, Antonio Loera was sentenced to 90 days in connection with the death of Steven Stayner.

Birth Data for Steven Stayner:

Birth date:	April 18, 1965		
Birth time:	02:29 AM PST		
Birth place:	Merced, CA		
Longitude:	120W29	Latitude:	37N18
Source:	Birth certificate		

Placements and degrees of Stayner's charts:
Natal Placements

Moon	27:53 Sco	09th	Neptune	19:12 Sco R	09th
Sun	28:10 Ari	02nd	Pluto	14:01 Vir R	07th
Mercury	12:20 Ari R	02nd	North Node	16:17 Gem R	04th
Venus	29:48 Ari	02nd	Vertex	11:21 Vir	07th
Mars	08:44 Vir R	07th	East Point	01:03 Pis	01st
Jupiter	29:05 Tau	03rd	Fortuna	18:25 Vir	07th
Saturn	13:25 Pis	01st	Midheaven	05:02 Sag	
Uranus	11:05 Vir R	07th	Ascendant	18:42 Aqu	

Eleventh house position is 26:57 Sag
Twelfth house position is 19:31 Cap
Second house position is 04:19 Ari
Third house position is 09:18 Tau

Critical Degrees: None
To be noted: Venus at 29 degrees of Aries and Jupiter at the ominous degree of 29 Taurus
Intercepted Signs: Virgo in the seventh house of one-to-one relationships and Pisces in the first house of self.

Stayner was abducted on December 4, 1972. The following are his planetary placements for the period of his abduction:

Progressed Placements for 1972

Moon	22:07Aqu	01st	Neptune	19:01 Sco R	09th
Sun	05:00 Tau	02nd	Pluto	13:54 Vir R	07th
Mercury	12:12 Ari	02nd	North Node	15:55 Gem R	04th
Venus	08:27 Tau	02nd	Vertex	16:22 Vir	07th
Mars	08:54 Vir	07th	East Point	08:44 Pis	01st
Jupiter	00:38 Gem	03rd	Fortuna	16:11 Sag	10th
Saturn	14:05 Pis	01st	Midheaven	11:52 Sag	11th
Uranus	10:56 Vir R	07th	Ascendant	29:03 Aqu	01st

Critical Degrees: None

To be noted: Jupiter at 00 degrees of Gemini in the natal third house of everyday surroundings, and North Node at 29 degrees of Aquarius in the natal first house of self.

Transiting Placements for December 4, 1972

Moon	27:06 Sco	09th	Uranus	21:46 Lib	08th
Sun	12:23 Sag	10th	Neptune	05:15 Sag	10th
Mercury	26:23 Sco R	09th	Pluto	04:07 Lib	07th
Venus	12:12 Sco	09th	North Node	18:42 Cap R	11th
Mars	12:19 Sco	09th	Vertex	15:22 Tau	03rd
Jupiter	11:35 Cap	11th	East Point	21:45 Lib	08th
Saturn	17:29 Gem R	04th	Fortuna	01:09 Lib	07th

Critical Degrees: Saturn
To be noted: None

Solar Return Placements for 1972 (Merced, CA)

Moon	27:36 Gem	08th	Neptune	04:48 Sag R	02nd
Sun	28:10 Ari	06th	Pluto	29:59 Sco R	11th
Mercury	04:21 Ari	05th	North Node	00:54 Aqu R	03rd
Venus	13:31 Gem	08th	Vertex	09:52 Gem	08th
Mars	14:15 Gem	08th	East Point	08:41 Sco	01st
Jupiter	08:15 Cap	03rd	Fortuna	29:02 Sag	02nd
Saturn	04:35 Gem	08th	Midheaven	03:59 Leo	
Uranus	15:47 Lib R	12th	Ascendant	29:36 Lib	

Eleventh house position is 06:53 Vir
Twelfth house position is 05:38 Lib
Second house position is 28:00 Sco
Third house position is 29:53 Sag

Critical Degrees: Saturn, Neptune
To be noted: Pluto at 29 degrees of Scorpio, North Node at 00 degrees of Aquarius, Fortuna or Misfortuna at 29 degrees of Sagittarius, and the Ascendant at 29:36 degrees of Libra
Intercepted Signs: Cancer intercepted in the ninth house of abstract mind and Capricorn intercepted in the third house of everyday environment

Stayner walked into the police station with a five-year old boy. The following are his planetary placements for the period of his returning home:

Progressed Placements for 1979

Moon	24:05 Tau	03rd	Neptune	18:50 Sco R	09th
Sun	11:48 Tau	03rd	Pluto	13:49 Vir R	07th
Mercury	15:49 Ari	02nd	North Node	15:33 Gem R	04th

Venus	17:05 Tau	03rd	Vertex	21:27 Vir	07th
Mars	09:38 Vir	07th	East Point	16:36 Pis	01st
Jupiter	02:12 Gem	03rd	Fortuna	22:30 Pis	01st
Saturn	14:41 Pis	01st	Midheaven	18:40 Sag	
Uranus	10:49 Vir R	07th	Ascendant	10:12 Pis	

Critical Degrees: None
To be noted: None

Transiting Placements for March 1, 1980

Moon	06:15 Vir	07th	Uranus	25:34 Sco	09th
Sun	11:00 Pis	01st	Neptune	22:31 Sag	10th
Mercury	20:01 Pis R	01st	Pluto	21:24 Lib R	08th
Venus	23:55 Ari	02nd	North Node	28:42 Lib R	07th
Mars	03:44 Vir R	07th	Vertex	08:59 Leo	06th
Jupiter	04:33 Vir R	07th	East Point	14:53 Cap	11th
Saturn	24:38 Vir R	07th	Fortuna	22:23 Gem	04th

Critical Degrees: Jupiter, Vertex approaching the critical degree of 9 Leo
To be noted: None

Solar Return Placements for 1979 (Manchester, CA)

Moon	15:45 Cap	06th	Neptune	20:19 Sag R	05th
Sun	28:10 Ari	10th	Pluto	17:33 Lib R	03rd
Mercury	01:02 Ari	09th	North Node	15:31 Vir R	02nd
Venus	24:46 Pis	09th	Vertex	17:16 Sag	05th
Mars	09:08 Ari	09th	East Point	16:09 Can	12th
Jupiter	29:53 Can	12th	Fortuna	19:40 Ari	10th
Saturn	07:27 Vir R	02nd	Midheaven	18:59 Ari	
Uranus	19:52 Sco R	04th	Ascendant	02:05 Leo	

Eleventh house position is 25:17 Tau
Twelfth house position is 01:06 Can
Second house position is 22:52 Leo
Third house position is 17:50 Vir

Critical Degrees: Vertex
To be noted: Jupiter at 29 degrees of Cancer
Intercepted Signs: Gemini intercepted in the eleventh house of friends, hopes and wishes and Sagittarius intercepted in the fifth house of children, love relationships and creativity.

Stayner died in a hit-and-run accident in 1989. The following are his planetary placements for the period when he died:

Progressed Placements for 1989

Moon	14:48 Lib	08th	Neptune	18:34 Sco R	09th
Sun	21:29 Tau	03rd	Pluto	13:43 Vir R	07th
Mercury	25:55 Ari	02nd	No. Node	15:01 Gem R	03rd
Venus	29:24 Tau	03rd	Vertex	28:45 Vir	07th
Mars	11:33 Vir	07th	East Point	28:02 Pis	01st
Jupiter	04:30 Gem	03rd	Fortuna	20:23 Leo	07th
Saturn	15:28 Pis	01st	Midheaven	28:20 Sag	11th
Uranus	10:43 Vir R	07th	Ascendant	27:03 Pis	01st

Critical Degrees: The Sun, Mercury approaching the critical degree of 26 Aries, Jupiter
To be noted: Venus at 29 degrees of Taurus in the third house of everyday environment

Transiting Placements for September 16, 1989

Moon	06:49 Ari	02nd	Uranus	01:21 Cap	11th
Sun	23:32 Vir	07th	Neptune	09:37 Cap R	11th
Mercury	09:25 Lib R	08th	Pluto	13:12 Sco	09th
Venus	04:34 Sco	08th	North Node	24:06 Aqu R	01st
Mars	27:57 Vir	07th	Vertex	27:40 Sag	11th
Jupiter	08:04 Can	05th	East Point	29:56 Can	06th
Saturn	07:19 Cap	11th	Fortuna	26:19 Aqu	01st

Critical Degrees: None
To be noted: The East Point at 29:56 Cancer

Solar Return Placements for 1989 (Merced, CA)

Moon	26:39 Vir	10th	Neptune	12:23 Cap R	02nd
Sun	28:10 Ari	05th	Pluto	14:15 Sco R	11th
Mercury	12:36 Tau	05th	North Node	02:06 Pis R	03rd
Venus	01:35 Tau	05th	Vertex	21:16 Can	08th
Mars	23:14 Gem	07th	East Point	22:05 Sag	01st
Jupiter	06:50 Gem	07th	Fortuna	04:28 Tau	05th
Saturn	13:55 Cap	02nd	Midheaven	20:38 Vir	
Uranus	05:18 Cap R	01st	Ascendant	05:59 Sag	

Eleventh house position is 20:53 Lib
Twelfth house position is 15:13 Sco

Second house position is 7:59 Cap
Third house position is 14:35 Aqu

Critical Degrees: Saturn
To be noted: None
Intercepted Signs: None

Lindsay Bay
Have You Seen My Missing Child?

The poster displayed the anguish of a Connecticut mother in December 1987. It read "Have You Seen My Missing Child." Her two year-old daughter, Lindsay Bay, had been kidnapped by her estranged husband. Enfield, CT police and FBI investigators were hindered in their work as they sorted through the cluster of clues left by the girl's father.

After Rhonda and Wayne Bay separated in June 1987, a restraining order was issued preventing the father from seeing Lindsay. However, Mrs. Bay felt he was being unfairly deprived of seeing his daughter and began giving him visitation rights. The couple divorced in November 1987.

In what was considered to be an act of vengeance, Wayne Bay abducted his daughter after her weekend visit with him on December 11, 1987. When he failed to return Lindsay, Mrs. Bay became concerned and called in police investigators. When the investigators went to Bay's apartment, they found a letter written to his ex-wife in which he told her that she would never see her daughter again. A milk carton sat on his table with two photographs of Lindsay pasted over the pictures of other missing children. Beneath Lindsay's pictures he had printed:

Jane Doe
000 Gone Street
Where, Earth 00000

The clues left behind by Bay baffled the investigators for several months after Lindsay's abduction. Before the abduction, Bay had gone out of his way to tell people he was leaving for either Australia or New Zealand. Investigators described Bay as a "con artist" and "chronic liar." The police found it difficult to separate the false clues from real clues. Although Bay had sold practically everything he owned, the police found his car in the long-term parking at the local airport. A cassette in the tape deck played the sound of an airplane taking off from a runway. The police then obtained a warrant for Bay's arrest charging him with second-degree kidnapping and custodial interference.

Investigators had two theories regarding Lindsay's abduction. Either Bay had planned the abduction for several months, or he had decided to flee the Connecticut area in order to avoid being charged with several unrelated charges at the time of the abduction. Bay was adept at conning others to his advantage. The tape left in Bay's car indicated that he and Lindsay were on their way to New Zealand, but this would have required him to secure passports, transportation and an ample supply of medicine for Lindsay's chronic ear infections. However, transiting Uranus was in conjunction with Bay's fifth house cusp on December 17, 1987, indicating a sudden, unexpected decision in connection with children.

In this case, a psychic was able to provide other clues to help in solving this kidnapping. On March 20, 1988, Mrs. Bay met with a group of four astrologers/psychics in upstate New York. One of the psychics who had just moved to New York from the Midwest drew a radio tower similar to one that he had seen in Seattle. He stated, "I don't know where this tower is, but it is near water. Lindsay is near this tower." Others recognized that the tower was similar to those leftover from the World's Fair in Montreal. The direction of "going north" and Montreal had already been mentioned before he made the drawing. In addition, Mrs. Bay was told that she would have her daughter home before May 1.

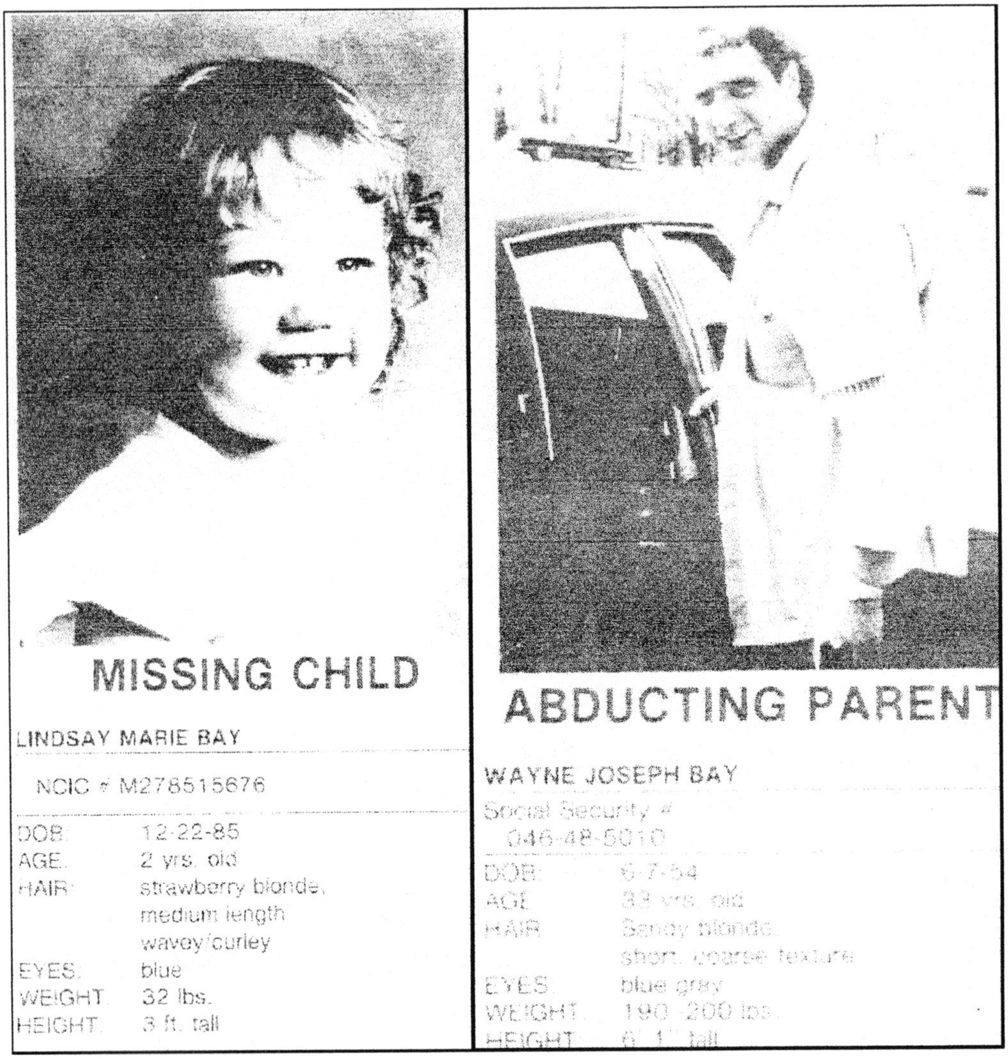

Illustration 17 - Lindsay Bay's Missing Child Poster with abducting parent (Bay)

Mrs. Bay took a tape of the meeting back to Connecticut with her. The next time she saw the FBI investigator, she gave him the copy. He slyly said to her, "I hope this didn't cost you much." She later told us that when he listened to the tape on his way home a couple of nights later, it changed his thinking on the case--and the investigation. Investigators were able to locate Bay in a town on the outskirts of Montreal. Mrs. Bay and her father went to Canada, taking all the records from the Enfield Police Department and the FBI with them. Through a lawyer they had hired in Montreal, they obtained a removal order so that they could take custody of Lindsay. The police accompanied them to Bay's residence, and he relinquished Lindsay with no fight.

Lindsay Bay's Natal Chart

Lindsay Bay has a stellium in the sign of Sagittarius in the fifth and sixth houses. It is comprised of Saturn at the critical degree of 04:11 Sagittarius, Fortuna or Misfortuna at 04:57 Sagittarius, the Vertex at 08:53 Sagittarius, Mercury at 10:31 Sagittarius, Uranus at 18:58 Sagittarius, and Venus in the sixth house at 24:26 Sagittarius. Saturn trines the Midheaven at 04:05 Aries.

Mars at 05:02 Scorpio in the fourth house trines the Midheaven at 04:05 Aries. Mars also conjuncts both the South Node at 06:18 Scorpio R and Pluto at 06:41 Scorpio in the fourth house of home and personal areas. Mars at 05:02 Scorpio inconjuncts the Midheaven at 04:05 Aries. The Sun conjuncts Neptune at 03:15 Capricorn in the sixth house of work and service, and Neptune squares the Midheaven at 04:05 Aries.

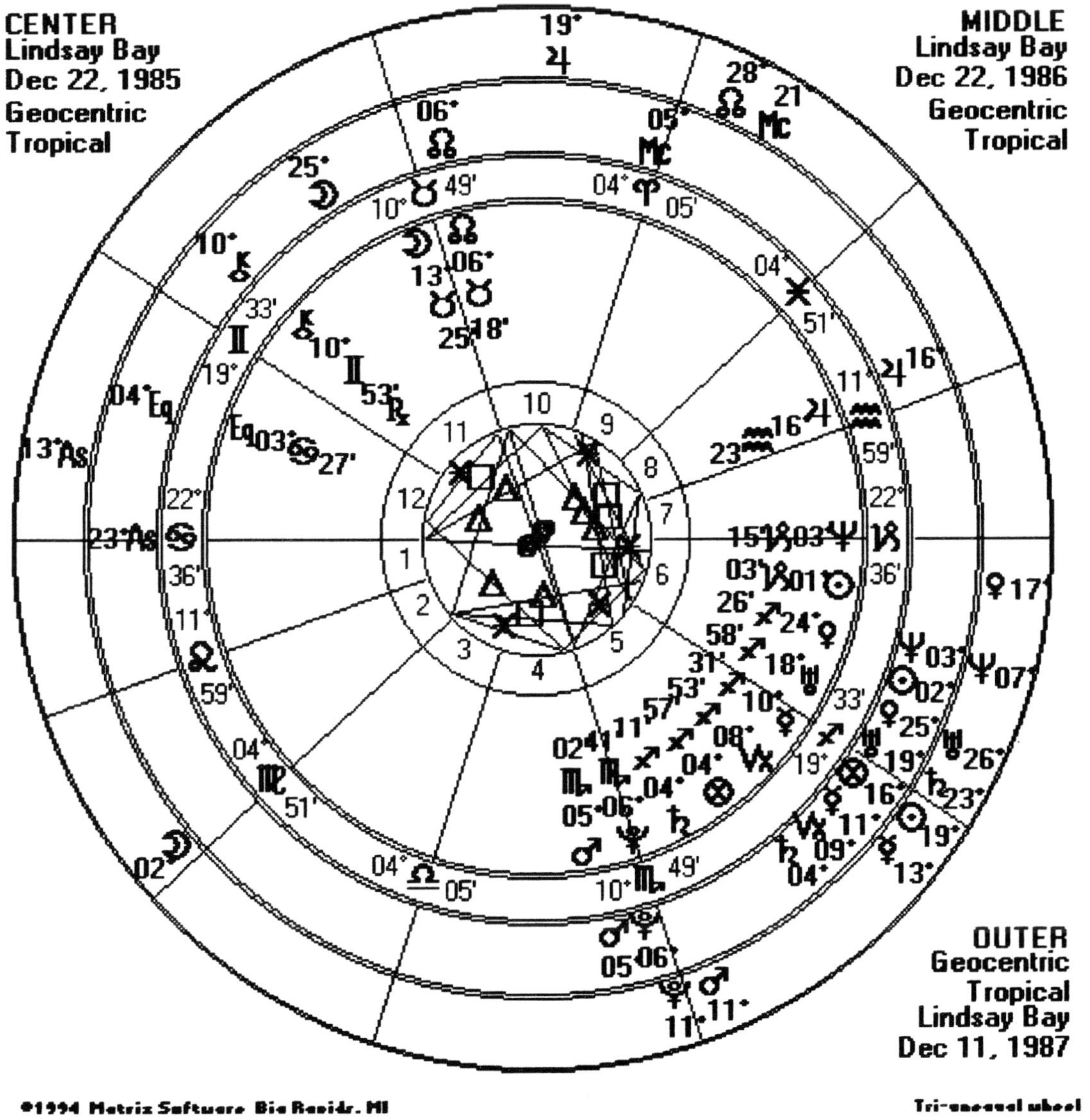

Chart 26 - Lindsay Bay's Three-ring Chart with Natal, Progressed Chart for 1986 and Transits for December 11, 1987

The Progressed Chart for December 22, 1986

Lindsay Bay's progressed chart also has a stellium Sagittarius in the natal fifth and sixth houses. It is comprised of Saturn at the critical degree of 04:18 Sagittarius, the Vertex at 09:37 Sagittarius, Mercury at 11:48 Sagittarius, Fortuna or Misfortuna at 16:28 Sagittarius, Uranus at 19:02 Sagittarius, all in the natal fifth house, and Venus at 25:41 Sagittarius in the natal sixth house.

Progressed Venus at 25:41 Sagittarius in the natal sixth house inconjuncts the progressed Moon at 25:11 Taurus in the natal eleventh house. Progressed Mars at 05:39 Scorpio, South Node at 06:15 Scorpio R, and Pluto at 06:43 Scorpio, all in the natal fourth house of home and personal matters, oppose the North Node at 06:15 Taurus R in the natal tenth house of public recognition. Progressed Jupiter at 16:35 Aquarius in the natal eighth house sextiles natal Fortuna or Misfortuna at 16:28 Sagittarius in the natal fifth house.

Lindsay Bay's Transits for December 11, 1987

Transiting Jupiter at 19:47 Aries R in the natal tenth house of public recognition and authority figures approaches a trine with natal Uranus at 18:58 Sagittarius in the natal fifth house. This aspect is exact on December 14. Transiting Saturn at 23:06 Sagittarius in the natal sixth house of health approaches a conjunction with natal Venus at 24:26 Sagittarius. This aspect is exact on December 22. Transiting Neptune at 07:02 Capricorn in the sixth house is separating from a sextile with natal Pluto at 06:41 Scorpio in the fourth house of home and personal matters.

An Annular Solar Eclipse on September 23, 1987 occurred at 29:33 Virgo. The eclipse falls in Lindsay Bay's third house and can indicate changes in her everyday environment and short-distance travels.

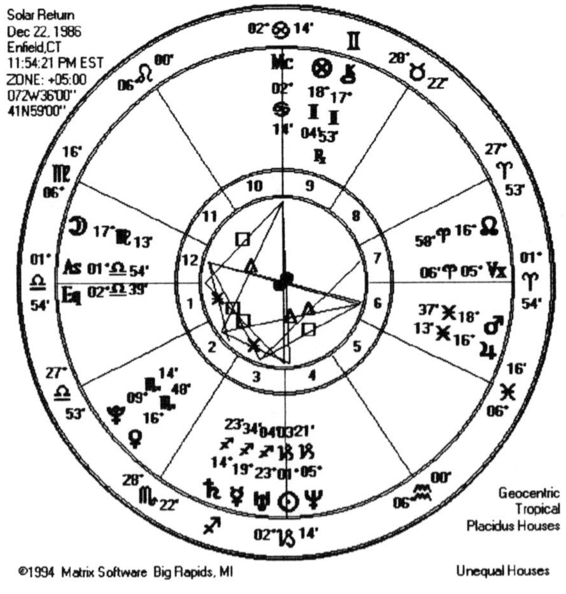

Chart 27 - Lindsay Bay's Solar Return Chart for 1986

The Solar Return Chart for December 1986 (Enfield, CT)

Lindsay Bay's solar return chart has both a yod and a grand cross. The yod is formed by solar Venus at 16:48 Scorpio in the solar second house of finances and values, the North Node at 16:58 Aries R in the solar seventh house of one-to-one relationships, and the Moon at 17:13 Virgo in the solar secret enemies and karma.

The mutable grand cross is formed by solar Mercury at 19:34 Sagittarius in the solar third house of everyday environment, Mars at 18:37 Pisces in the sixth house of health, Fortuna or Misfortuna at 18:04 Gemini in the ninth house of abstract mind and the Moon at 17:13 Virgo in the twelfth house.

The solar Moon is at the critical degree of 17:13 Virgo in the twelfth house and solar Pluto is at the critical degree of 09:14 Scorpio in the second house. The solar Moon sextiles Venus at 16:48 Scorpio in the second house and opposes both Mars at 18:37 Pisces and Jupiter at 16:13 Pisces in the sixth house. The solar Moon also inconjuncts the North Node at 16:58 Aries R in the seventh house. Pluto forms no close aspects.

Gemini is intercepted in the solar ninth house, and Sagittarius is intercepted in the solar third house.

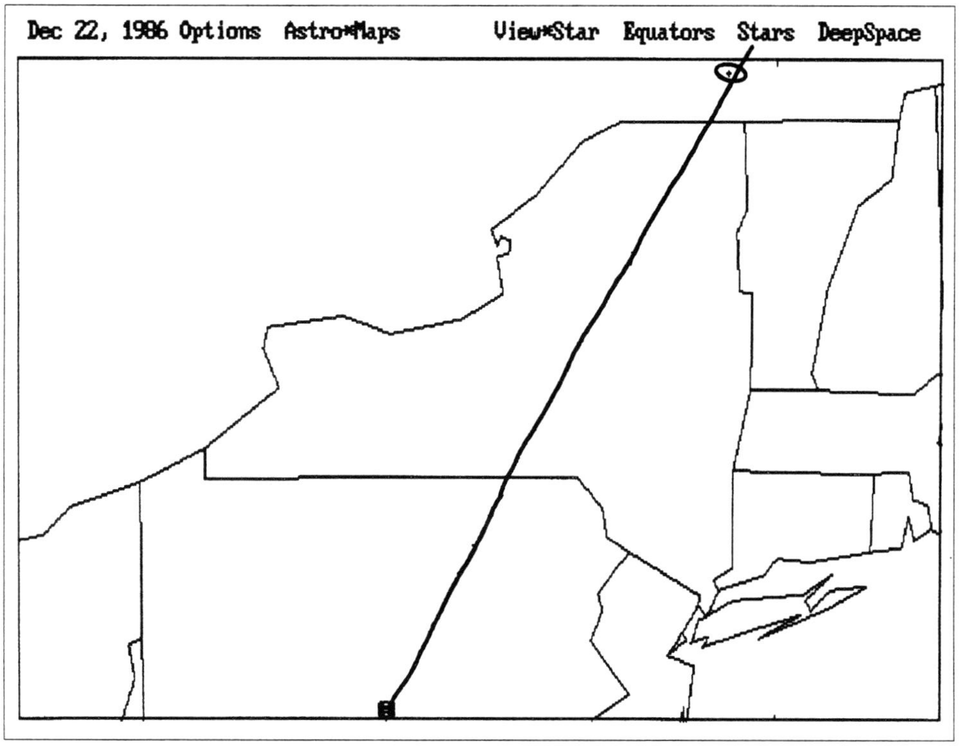

Map 6 - Lindsay Bay's Progressed Cyclo*Carto*Graphy Map for 1986

Astro*Carto*Graphy

Lindsay Bay's progressed Cyclo*Carto*Graphy map for 1986 has Mercury running through Montreal, Canada. This is an interesting placement due to Mercury's relationship to the third house of short-distance travel and everyday environment.

Wayne Bay's Transits for December 11, 1987

It was mentioned earlier that transiting Uranus at 26:26 Sagittarius was approaching a conjunction with Bay's fifth house cusp at 26:50 Sagittarius on December 17, 1987, indicating a sudden, unexpected decision in connection with children. The transiting Moon at 28:48 Leo in the first house of self is approaching the decisive 29 degrees.

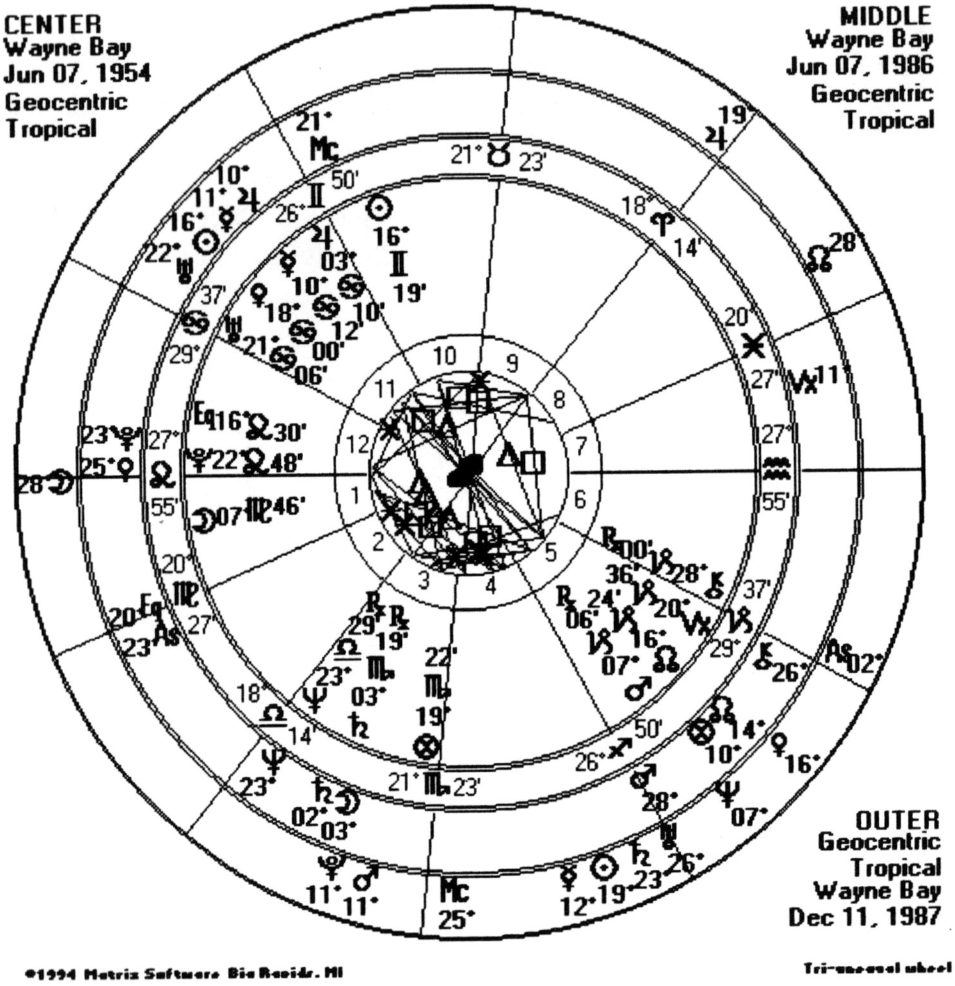

Chart 28 - Bay's Father's Three-ring Chart with Natal, Progressed Chart for 1986 and Transits for December 11, 1987

Transiting Neptune at 07:01 Capricorn is approaching a conjunction with natal Mars at 07:06 Capricorn in the fifth house of children. In *Predictive Astrology*, Frances Sakoian and Louis Acker interpret this conjunction as follows: "Secretive or covert action characterize this transit...The individual's subconscious desires and feelings may be stimulated at this time with the result that subconscious forces may bring uncontrolled impulses into the open."

An Annular Solar Eclipse on September 23, 1987 occurred at 29:33 Virgo. The eclipse falls in Bay's second house and indicates changes in finances and values.

Mother's Transits for December 11, 1987

Transiting Mercury at 12:08 Sagittarius in the natal fourth house of home and personal matters squares natal Pluto at 12:32 Virgo in the first house of self. Transiting Venus at 16:29 Capricorn in the fifth house of children squares natal Jupiter at 16:25 Aries R in the eighth house of shared resources. Transiting Saturn at 23:02 Sagittarius in the fourth house squares natal Mercury at 23:23 Virgo R in the first house.

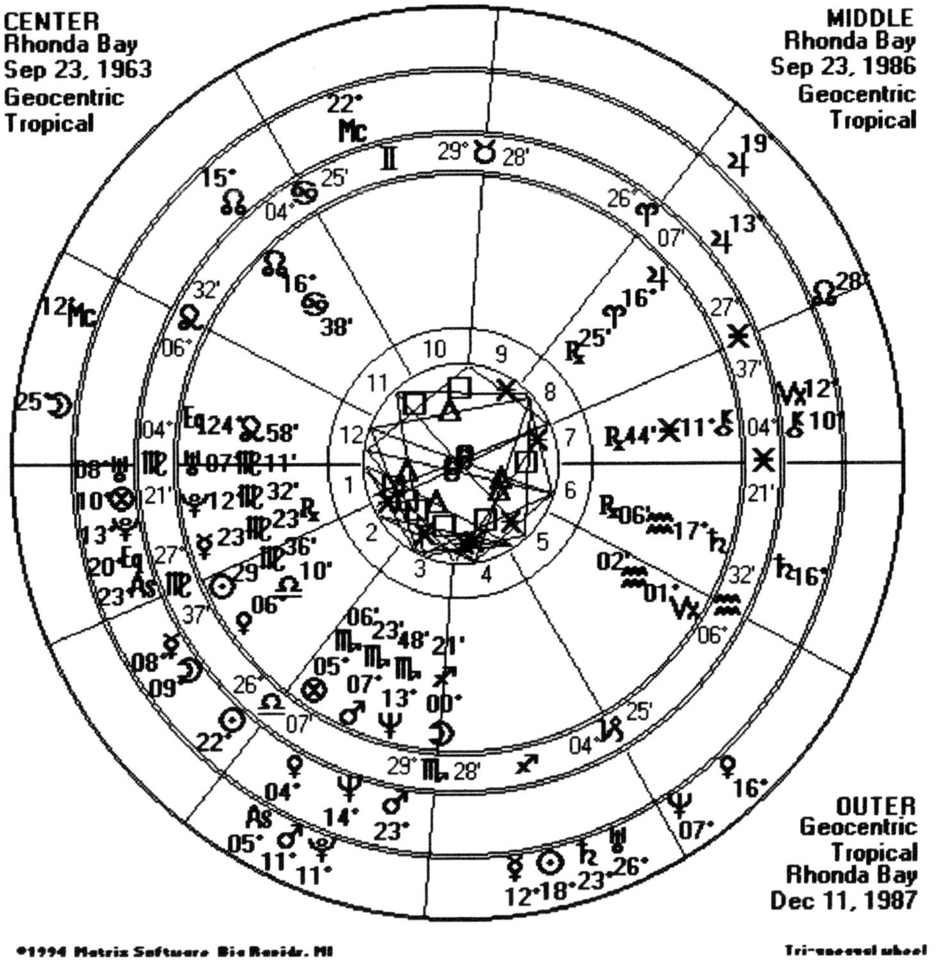

Chart 29 - Bay's Mother's Three-ring Chart with Natal, Progressed Chart for 1986 and Transits for December 11, 1987

An Annular Solar Eclipse on September 23, 1987 occurred at 29:33 Virgo. The eclipse falls in the mother's second house and indicates changes in finances and values. The eclipse is also conjunct the mother's natal Sun at 29:26 Virgo.

Mother's Transits for May 1, 1988

It had been predicted that Lindsay would be home before May 1, 1988. She was returned to Mrs. Bay on April 30, 1988. On May 2, transiting Jupiter at 12:27 Taurus in the ninth house of the law trines natal Pluto at 12:32 Virgo in the first house of self.

In Conclusion

This is one case with a happy ending, and the following note was received from Lindsay's mother:

> ***"I'd like to thank you so much for everything you and your friends did for us. Lindsay came home on Saturday, April 30th..."***

Birth Data for Lindsay Bay:

Birth date:	December 22, 1985		
Birth time:	06:00 PM/EST		
Birth place:	Hartford, CT		
Longitude:	072W41	Latitude:	41N46
Source:	Her mother		

Birth Data for Lindsay Bay's Parents:

Mother:	Rhonda Bay		
Birth date:	September 23, 1963		
Birth time:	04:32 AM/EDT		
Birth place:	Williamantic, CT		
Longitude:	072W13	Latitude:	41N43
Source:	From self		

Father:	Wayne Bay		
Birth date:	June 7, 1954		
Birth time:	11:04 AM/EST		
Birth place:	Manchester, CT		
Longitude:	072W31	Latitude:	41N47
Source:	From wife		

Placements and degrees of Lindsay's charts:
Natal Placements

Moon	13:25 Tau	11th	Neptune	03:15 Cap	06th
Sun	01:03 Cap	06th	Pluto	06:41 Sco	04th
Mercury	10:31 Sag	05th	North Node	06:18 Tau R	10th
Venus	24:26 Sag	06th	Vertex	08:53 Sag	05th
Mars	05:02 Sco	04th	East Point	03:27 Can	12th
Jupiter	16:23 Aqu	08th	Fortuna	04:57 Sag	05th
Saturn	04:11 Sag	05th	Midheaven	04:05 Ari	
Uranus	18:58 Sag	05th	Ascendant	22:36 Can	

Eleventh house position is 10:49 Tau
Twelfth house position is 19:33 Gem
Second house position is 11:59 Leo
Third house position is 04:51 Vir

Critical Degrees: Saturn
To be noted: None
Intercepted Signs: None

Progressed Placements for 1986

Moon	25:11 Tau	11th	Neptune	03:17 Cap	06th
Sun	02:04 Cap	06th	Pluto	06:43 Sco	04th
Mercury	11:48 Sag	05th	North Node	06:15 Tau R	10th
Venus	25:41 Sag	06th	Vertex	09:37 Sag	05th
Mars	05:39 Sco	04th	East Point	04:18 Can	12th
Jupiter	16:35 Aqu	08th	Fortuna	16:28 Sag	05th
Saturn	04:18 Sag	05th	Midheaven	05:07 Ari	
Uranus	19:02 Sag	05th	Ascendant	23:21 Can	

Critical Degrees: Saturn
To be noted: None

Transiting Placements for December 11, 1987

Moon	02:42 Vir	02nd	Uranus	26:27 Sag	06th
Sun	19:22 Sag	05th	Neptune	07:02 Cap	06th
Mercury	13:04 Sag	05th	Pluto	11:24 Sco	05th
Venus	17:14 Cap	06th	North Node	28:14 Pis R	09th
Mars	11:41 Sco	05th	Vertex	00:16 Sag	05th
Jupiter	19:47 Ari R	10th	East Point	23:03 Gem	12th
Saturn	23:06 Sag	06th	Fortuna	26:39 Pis	09th

Critical Degrees: None
To be noted: Vertex at 00 degrees of Sagittarius

Solar Return Placements for 1986 (Enfield, CT)

Moon	17:13 Vir	12th	Neptune	05:21 Cap	04th
Sun	01:03 Cap	03rd	Pluto	09:14 Sco	02nd
Mercury	19:34 Sag	03rd	North Node	16:58 Ari R	07th
Venus	16:48 Sco	02nd	Vertex	05:06 Ari	07th
Mars	18:37 Pis	06th	East Point	02:39 Lib	01st
Jupiter	16:13 Pis	06th	Fortuna	18:04 Gem	09th
]Saturn	14:23 Sag	03rd	Midheaven	02:14 Can	
Uranus	23:04 Sag	03rd	Ascendant	01:54 Lib	

Eleventh house position is 06:00 Leo
Twelfth house position is 06:16 Vir
Second house position is 27:53 Lib
Third house position is 28:22 Sco

Critical Degrees: Moon and Pluto
To be noted: None
Intercepted Signs: Gemini is intercepted in the ninth house of abstract mind, and Sagittarius is intercepted in the third house of everyday environment.

Placements and degrees of Father's charts:
Natal Placements

Moon	07:46 Vir	01st	Neptune	23:29 Lib R	03rd
Sun	16:19 Gem	10th	Pluto	22:48 Leo	12th
Mercury	10:12 Can	11th	North Node	16:24 Cap R	05th
Venus	18:00 Can	11th	Vertex	20:36 Cap	05th
Mars	07:06 Cap R	05th	East Point	16:30 Leo	12th
Jupiter	03:10 Can	11th	Fortuna	19:22 Sco	03rd
Saturn	03:19 Sco R	03rd	Midheaven	21:23 Tau	
Uranus	21:06 Can	11th	Ascendant	27:55 Leo	

Eleventh house position is 26:50 Gem
Twelfth house position is 29:37 Can
Second house position is 20:27 Vir
Third house position is 18:14 Lib

Critical Degrees: Midheaven
To be noted: None
Intercepted Signs: None

Progressed Placements for 1986

Moon	03:58 Sco	03rd	Neptune	23:17 Lib	03rd
Sun	16:51 Can	11th	Pluto	23:29 Leo	12th
Mercury	11:34 Can R	11th	North Node	14:42 Cap R	05th
Venus	25:41 Leo	12th	Vertex	11:41 Pis	07th
Mars	28:16 Sag R	05th	East Point	20:26 Vir	01st
Jupiter	10:24 Can	11th	Fortuna	10:12 Cap	05th
Saturn	02:39 Sco	03rd	Midheaven	21:56 Gem	
Uranus	22:57 Can	11th	Ascendant	23:06 Vir	

Critical Degrees: None
To be noted: None

Transiting Placements for December 11, 1987

Moon	28:48 Leo	01st	Uranus	26:26 Sag	04th
Sun	19:02 Sag	04th	Neptune	07:01 Cap	05th

Mercury	12:33 Sag	04th	Pluto	11:23 Sco	03rd
Venus	16:49 Cap	05th	North Node	28:15 Pis R	08th
Mars	11:28 Sco	03rd	Vertex	03:08 Vir	01st
Jupiter	19:47 Ari R	09th	East Point	20:53 Aqu	06th
Saturn	23:04 Sag	04th	Fortuna	12:12 Lib	02nd

Critical Degrees:
To be noted: Moon approaching 29 degrees of Leo, Uranus, the planet of sudden, unexpected situations at 26:26 Sagittarius and approaching the cusp of the fifth house of children at 26:50 Sagittarius, and the East Point approaching 21 degrees of Aquarius,

Solar Return Placements for 1986 (Enfield, CT)

Moon	13:57 Gem	01st	Neptune	04:56 Cap R	08th
Sun	16:19 Gem	01st	Pluto	04:55 Sco R	06th
Mercury	03:18 Can	02nd	North Node	27:29 Ari R	12th
Venus	20:03 Can	03rd	Vertex	02:38 Sco	06th
Mars	23:06 Can	09th	East Point	17:01 Tau	12th
Jupiter	20:54 Ari	11th	Fortuna	04:46 Gem	12th
Saturn	05:39 Sag R	06th	Midheaven	12:05 Aqu	
Uranus	20:33 Sag R	07th	Ascendant	07:08 Gem	

Eleventh house position is 10:48 Pis
Twelfth house position is 21:10 Ari
Second house position is 29:53 Gem
Third house position is 20:00 Can

Critical Degrees: Fortuna or Misfortuna
To be noted: None
Intercepted Signs: Taurus intercepted in the twelfth house and Scorpio intercepted in the sixth house

HOMICIDES

Betty Conley
A Family Asks Why

Betty Conley took a job at the Xtra Mart in the small upstate New York town of Charlton. She agreed to work the 11:00 PM to 6:00 AM shift for eight weeks in order to save the money that she needed to return to college that fall. She had been there for six weeks and was working alone on July 8, 1993. At approximately 02:10 AM/EDT that night she was shot in the back of the head. Although about $100 was taken from the register, robbery may not have been the reason for Conley's murder. A cash box which sat on the shelf under the register was left untouched. The longitude and latitude for Charlton, NY are 73W58 and 42N56.

Picture 6 - Betty Conley

Conley was married and had two teenage children. She also worked as a chef at a leading restaurant near her home and held another part-time job at a friend's nursery and garden shop. She was anxiously awaiting an acceptance letter from a college where she planned to study horticulture. Conley's acceptance letter came on the day she was murdered.

Conley's Natal Chart

We find several close aspects in Conley's natal chart in addition to a yod. The natal Sun at 17:18 Aquarius in the fourth house of the emotional and physical home sextiles natal Mars at 15:45 Sagittarius in the third house of everyday environment. The natal Moon at 24:55 Sagittarius in the third house squares natal Venus at 24:57 Pisces in the sixth house of work and service to others.

Mercury at 27:43 Capricorn R in the fourth house of home and personal matters inconjuncts both Jupiter at 27:44 Leo R and Pluto at 27:37 Leo R in the eleventh house of hopes and wishes. Venus at 24:57 Pisces in the sixth house inconjuncts both Jupiter at 27:44 Leo R and Pluto at 27:37 Leo R in the eleventh house.

Jupiter at 27:44 Leo R conjuncts Pluto at 27:37 Leo R in the eleventh house of hopes and wishes. Saturn at 01:51 Sagittarius in the second house of finances and values trines Uranus at 29:35 Cancer R in the tenth house of career and public recognition.

The natal yod is comprised of Mercury at 27:43 R in the fourth house of home and personal matters, Venus at 24:57 Pisces in the sixth house of work and service to others, and Jupiter at 27:44 Leo R and Pluto at 27:37 Leo R in the eleventh house of hopes and wishes.

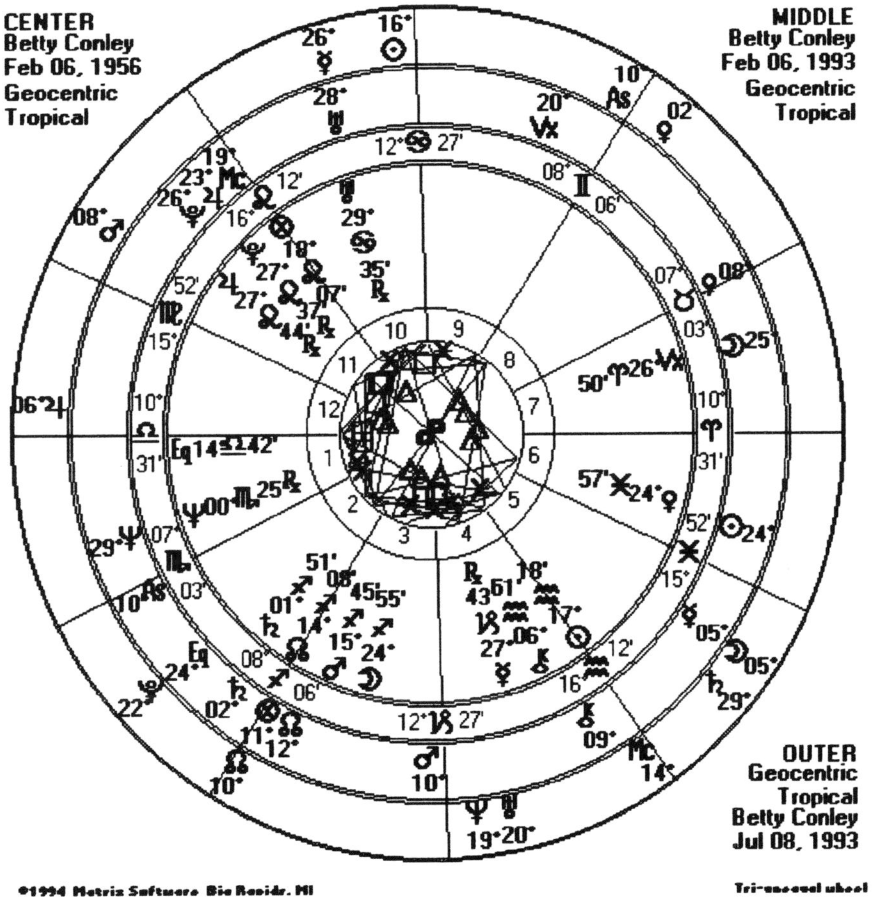

Chart 30 - Betty Conley's Three-ring Chart with Natal, Progressed Chart for 1993 and Transits for July 08, 1993

The Progressed Chart for 1993

The progressed Moon at 25:39 Aries in the natal seventh house of one-to-one relationships trines both progressed Jupiter at 23:11 Leo R and Pluto 26:44 Leo R in the natal eleventh house of friends and acquaintances. The progressed Sun at 24:30 Pisces in the natal sixth house of service and health inconjuncts both progressed Jupiter at 23:11 Leo R and Pluto 26:44 Leo R in the natal eleventh house.

Progressed Venus at 08:11 Taurus in the natal eighth house trines progressed Mars at 10:00 Capricorn in the third house of everyday environment, and Mars sextiles the natal Ascendant at 10:13 Scorpio. Progressed Uranus at 28:25 Cancer R in the natal tenth house of career and public recognition squares progressed Neptune at 29:56 Libra R and natal Venus at 00:25 Scorpio R both in the natal first house of self and environment.

Conley's Transits Interpreted for July 8, 1993

The transiting Sun at 16:08 Cancer in the natal tenth house of career and public recognition inconjuncts the natal Sun at 17:18 Aquarius in the natal fifth house of children and romantic relationships and natal Mars at 15:45 Sagittarius in the natal third house of everyday environment. Transiting Saturn at 29:42 Aquarius R in the natal fifth house inconjuncts natal Uranus at 29:35 Cancer R in the tenth house. Saturn retrogrades back to an exact aspect on July 10, 1997 and had previously occurred on May 10, 1997. Lois Rodden in *Modern Transits* interprets Saturn in an adverse aspect with Uranus as "Your work or security pattern is subject to erratic change."

Charlton store clerk murdered

By SYLVIA WOOD
Gazette Reporter

CHARLTON — Betty Conley's family had begged her to quit her overnight shift as a store clerk at the Xtra Mart on Route 67, just west of Ballston Spa.

On Thursday morning, just a few hours after Conley began her shift —

Removing temptation part of stores' defenses, Page A4

and only six weeks after she began working there — the family's worst fears came true.

The 37-year-old Providence mother of two was shot in the head with a single bullet about 2:10 a.m. during an apparent robbery of the convenience store, officials said.

She was found dead about an hour and a half later behind the counter by two customers.

Saratoga County Sheriff's Depart-

See MURDER, Page A4

JIM CASSIN Gazette Photographer

The Xtra Mart convenience store on Route 67 in Charlton was closed Thursday after store clerk Betty Conley, a 37-year-old mother of two, was gunned down in an apparent early-morning robbery. Police were searching for two vehicles in connection with the homicide.

Illustration 18 - Schenectady Gazette (Conley)

A Partial Solar Eclipse on May 21, 1993 occurred at 00:32 Gemini. The eclipse falls in Conley's eighth house and can indicate changes in shared resources.

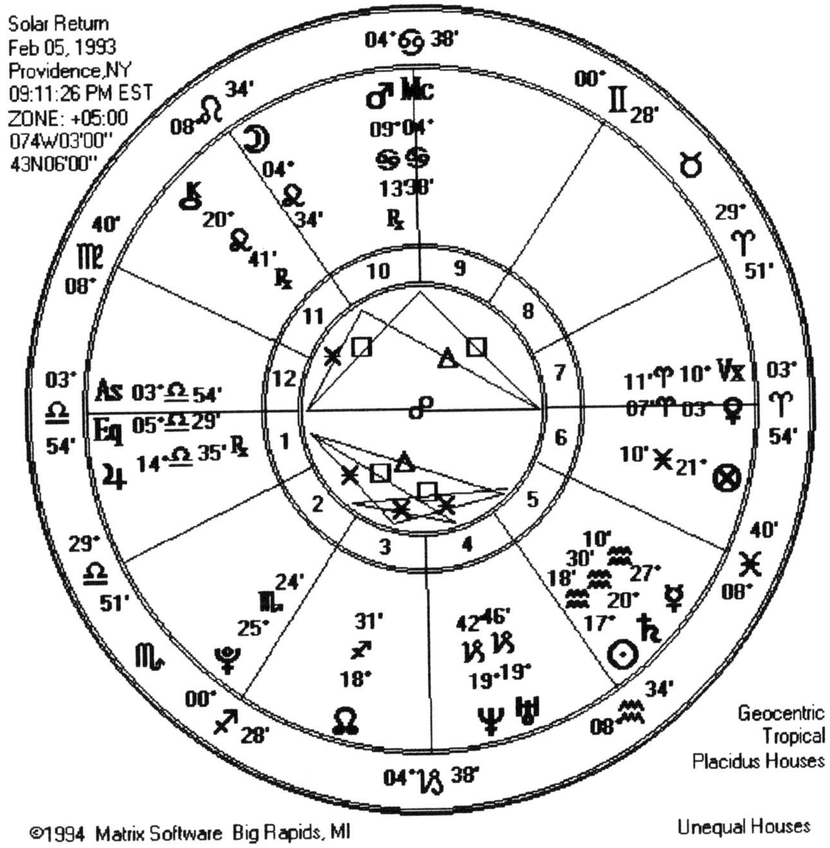

Chart 31 - Betty Conley's Solar Return Chart for 1993

The Solar Return Chart for 1993 (Providence, NY)

The solar Moon at 04:34 Leo in the tenth house of career and public recognition squares solar Venus at 03:07 Aries in the sixth house of work and service. The solar Sun at 17:18 Aquarius in the fifth house of children and romantic relationships sextiles the solar North Node at 18:31 Sagittarius R in the third house of everyday environment. Solar Mercury at 27:10 Aquarius squares solar Pluto at 25:24 Scorpio in the second house of personal finances and values.

Solar Venus at 03:07 Aries in the sixth house of work and service squares the solar Midheaven at 04:38 Cancer. Uranus at 19:46 closely conjuncts Pluto at 19:42 Capricorn in the fourth house of home and personal matters.

The Yod or Fingers of Fate

There is a yod in Conley's charts. It is formed by a third house natal Mars at 15:45 Sagittarius, a fifth house natal Sun at 17:18 Aquarius and a tenth house transiting Sun at 16:08 Cancer. Therefore, the yod includes:

The Yod

Third House	Natal Mars	15:45 Sagittarius
Fifth House	Natal Sun	17:18 Aquarius
Tenth House	Transiting Sun	16:08 Cancer

What follows is an interpretation of the yod in Betty's charts using a repetitive Basic Keywords from *The Rulership Book* by Rex E. Bills. They are separated into Victim and the Killer.

Keywords for Victim

Acquaintances, friends, powerful	Males, in general
Powerful persons/relatives	Everyday environment
One's employment, business	One's business life
Anger, arguments, conflict, disagreement	Temper, one with quick temper
Danger, dangerous occupation	Burglars, burglary, stealing
Crimes of violence	Gunman, gunshot wounds
Firearms, shooting, shootings	Back of the body, head, in general
Deaths by accident/violence	Deaths, sudden
Children, in general	Murder
Matters outside the home	One's fate

Keywords for Killer

Acquaintances	Everyday environment, neighbors
Neighborhood, counties, neighboring	Routine Coming and going
Trade, local	Acne, Blemishes, skin
Face, eruptions, pimples on the face	Complexion, reddish
Tattoos	Temper, one with quick temper
Nervous disorders	Introvert
Males, in general, men ages 25-35	Period of life, 24-34 (Sepherial)
Friends, powerful	Males, in general
Powerful persons/relatives	Relatives, male
Step-families, adopted children	Relatives, male
Army, armed forces, military, in general	Sportsmen, athletes, sports
Places near water, reservoirs	Water, occupations connected
Trailer houses	Parks, park keepers, caretakers
Public utilities, bathrooms	Telephones, booths, repairmen
Automobiles (mechanics, dealers, racers)	Motors, motorcycles, motormen
Taverns, owners and helpers	Cooks, diners, dining Rooms
Hotels, keepers, workers	Food, purveyors of food
Places where food is stored/prepared	Kitchens
Restaurants, owners and workers	Rooming houses
Dining, taverns, owners and helpers	

In Conclusion

For several months after Conley's murder, her husband remained the Saratoga County Sheriff Department's prime suspect. However, their evidence did not prove their suspicions to be correct. Although they followed up on other leads, the trails to new suspects were cold. Conley's murder is still unsolved.

Birth Data for Betty Conley:

Birth date:	February 06, 1956		
Birth time:	09:45 PM/EST		
Birth place:	Schenectady, NY		
Longitude:	073W56	Latitude:	42N49
Source:	From sister		

Placements and degrees of Betty Conley's charts:
Natal Placements

Moon	24:55 Sag	03rd	Neptune	00:25 Sco R	01st
Sun	17:18 Aqu	05th	Pluto	27:37 Leo R	11th
Mercury	27:43 Cap R	04th	North Node	14:08 Sag R	03rd
Venus	24:57 Pis	06th	Vertex	26:50 Ari	07th
Mars	15:45 Sag	03rd	East Point	14:42 Lib	01st
Jupiter	27:44 Leo R	11th	Fortuna	18:07 Leo	11th
Saturn	01:51 Sag	02nd	Midheaven	12:27 Can	
Uranus	29:35 Can R	10th	Ascendant	10:31 Lib	

Eleventh house position is 16:12 Leo
Twelfth house position is 15:52 Vir
Second house position is 07:03 Sco
Third house position is 08:06 Sag

Critical Degrees: Vertex
To be noted: Uranus at 29 degrees in the tenth house of recognition and employer and Neptune at 00 degrees in the first house of self.
Intercepted Signs: None

Progressed Placements for 1993

Moon	25:39 Ari	07th	Neptune	29:56 Lib R	01st
Sun	24:30 Pis	06th	Pluto	26:44 Leo R	11th
Mercury	05:47 Pis	05th	North Node	12:10 Sag R	03rd
Venus	08:11 Tau	08th	Vertex	20:19 Gem	09th
Mars	10:00 Cap	03rd	East Point	24:26 Sco	02nd
Jupiter	23:11 Leo R	11th	Fortuna	11:22 Sag	03rd
Saturn	02:48 Sag R	02nd	Midheaven	19:39 Leo	
Uranus	28:25 Can R	10th	Ascendant	10:13 Sco	

Critical Degrees: None
To be noted: Neptune at 29 degrees in the first house of self

Transiting Placements for July 8, 1993 (3:00 A.M./E.D.T.)

Moon	05:59 Pis	05th	Uranus	20:23 Cap R	04th
Sun	16:08 Can	10th	Neptune	19:52 Cap R	04th
Mercury	26:37 Can R	10th	Pluto	22:54 Sco R	02nd
Venus	02:28 Gem	08th	North Node	10:27 Sag R	03rd
Mars	08:45 Vir	11th	Vertex	04:58 Sco	01st
Jupiter	06:44 Lib	12th	East Point	19:46 Tau	08th
Saturn	29:42 Aqu R	05th	Fortuna	00:50 Aqu	04th

Critical Degrees: Mercury R and Mars
To be noted: Saturn at 29 degrees in the fifth house of children and love relationships and Fortuna or Misfortuna at 00 degrees in the fourth house of home and personal matters

Solar Return Placements for 1993 (Providence, NY)

Moon	04:34 Leo	10th	Neptune	19:42 Cap	04th
Sun	17:18 Aqu	05th	Pluto	25:24 Sco	02nd
Mercury	27:10 Aqu	05th	North Node	18:31 Sag R	03rd
Venus	03:07 Ari	06th	Vertex	10:11 Ari	07th
Mars	09:13 Can R	10th	East Point	05:29 Lib	01st
Jupiter	14:35 Lib R	01st	Fortuna	21:10 Pis	06th
Saturn	20:30 Aqu	05th	Midheaven	04:38 Can	
Uranus	19:46 Cap	04th	Ascendant	03:54 Lib	

Eleventh house position is 08:34 Leo
Twelfth house position is 08:40 Vir
Second house position is 29:51 Lib
Third house position is 00:28 Sag

Critical Degrees: None
To be noted: None
Intercepted Signs: Taurus intercepted in the eighth house of shared resources and Scorpio in the second house of personal finances and values

Eliza Hope Thomas
The "Yogurt Shop Murders"

On Friday evening, December 6, 1991, Eliza Thomas and one of her best friends, Jennifer Harbison, both 17, were working at the I Can't Believe It's Yogurt! shop in Austin, TX. The yogurt shop was a popular teenage hangout. Harbison's younger sister Sarah, 15, and her friend Amy Ayers, 13, had come to the yogurt shop to get a ride home when it closed at 11 P.M. Ayers was spending the night at the Harbisons. All of the parents thought their daughters were safe at the yogurt shop.

Illustration 19 - Western Sun - Eliza Thomas, Jennifer Harbison, Sarah Harbison and Amy Ayers

At 11:47 P.M., a fire at the yogurt shop was reported by an off-duty police officer who happened to be riding in the area. Upon investigation of the fire, the girl's bodies were found. All their bodies had been burned, three beyond recognition. Each girl's hands had been tied behind her and all four had been shot in the back of the head. The "Yogurt Shop Murders" stunned the residents of Austin, and although police received and investigated hundreds of leads involving hundreds of suspects, they were unable to make an immediate arrest in the case.

All four of the girls were members of the Lanier High School Future Farmers of America and had raised animals for stock-show competition. At the Austin-Travis County Livestock Show's Animal Auction in March 1992, Stoney, the hog that Eliza Thomas had raised, earned the title of 1992 Grand Champion when it was shown by her younger sister, Sonora. The 254-pound porker was bought for $6,000. Amy Ayer's 235-pound hog was shown by a friend who raised the hog after Ayer's death. It finished in fourth place and was sold for $2,750. Classmates continued to care for the Harbison's lambs after their deaths. Sarah Harbinson's lamb, Stella, earned sixth place, while Jennifer Harbinson's lamb, Casey, finished ninth. The lambs sold for $2,100 each and were returned to their caretakers after the auction.

On March 25, 1992, "48 Hours" featured the story of the "Yogurt Shop Murders." "America's Most Wanted" has also shown the story. Finally in October 1992, an indictment was issued for questioning in connection with the "Yogurt Shop Murders" for two Mexican nationals, Porfioro Villa Saavedra, also known as the "Terminator," and Alberto Jimenez Cortes. Police have been unable to locate another man, Ricardo

Hernandez, who lived with Saavedra and Cortez in Austin at the time of the murders and is also wanted for questioning. Saavedra, the key suspect, later confessed to the murders, but recanted and said that he was tortured by police to make him confess. Cortez, who implicated Saavedra also recanted.

Staff photo by Taylor Johnson

Nick Larsen, a former schoolmate of Eliza Thomas', stopped by the yogurt shop Sunday where his friend and three other teens were murdered this weekend. Mourners paid their respects throughout the day with cards, tears and silence.

Teens' violent deaths mourned

By Sylvia Martinez
American-Statesman Staff

Each of the four teen-age girls slain in a Northwest Austin yogurt store had been shot twice in the back of the head and three of the victims were burned

blaze at the I Can't Believe It's Yogurt store at 2949 W. Anderson Lane.

The teens died before the store was set ablaze, said Travis County Medical Examiner Robert Bayardo.

Police, who believe the fire was set in

Bayardo said there was little doubt about identities and he expected to confirm them through dental records.

Though Bayardo said he did not suspect sexual assault, police ordered rape examinations on the victims.

"We felt in this case it was definitely

Illustration 20 - The Austin American-Statesman
(Eliza Thomas, Jennifer Harbison, Sarah Harbison and Amy Ayers)

Eliza Thomas's chart will be used for this chapter because an accurate time has been taken from her birth certificate. She was one of the four teenage girls murdered at the yogurt shop at approximately 10:15 PM/CST. Thomas was a senior in high school and had started to work part-time at the yogurt shop in December 1990. The longitude and latitude for Austin, TX are 97W44 and 30N16.

Thomas's Natal Chart

In Thomas's natal chart, the Moon at 15:43 in the first house of self trines Mars at 15:33 Cancer in the fifth house of love relationships and creativity. Venus approaching the critical degree of 13 Aries is at 12:47 Aries in the first house and trines the Midheaven at 13:24 Sagittarius. Jupiter at 13:45 Pisces in the first house squares the Midheaven at 13:24 Sagittarius. Neptune at 08:37 Sagittarius R in the ninth house opposes Mercury 08:16 Gemini in the third house of everyday environment.

Thomas's natal Sun at 25:00 Taurus in the third house of everyday environment is conjunct Algol at 25:03 Taurus. Algol is said to be the most violent star and it is connected with strangulation, beheading or other violent injury to the head, throat and neck.

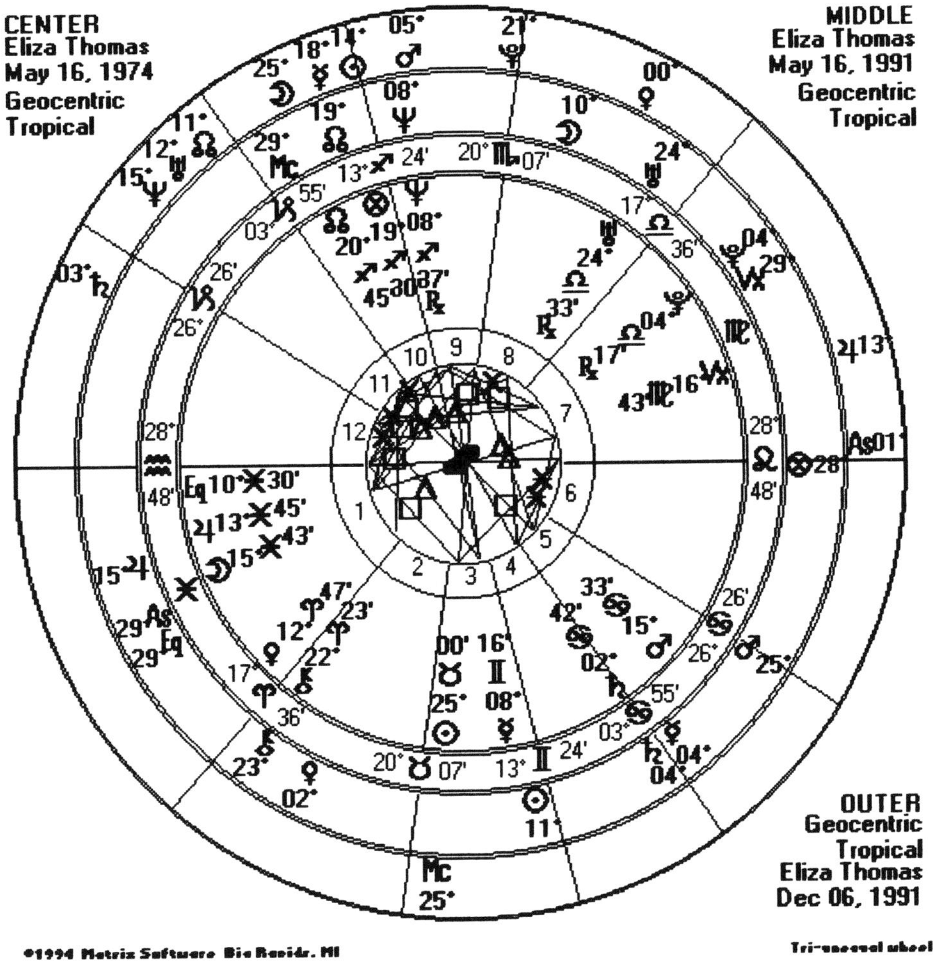

Chart 32 - Eliza Thomas's Three-ring Chart with Natal, Progressed Chart for 1991 and Transits for December 6, 1991

The Progressed Chart for 1991

All four angles of Thomas's progressed chart are at 29 degrees:

Progressed Midheaven	29:45 Sagittarius
Progressed Ascendant	29:30 Pisces

In addition, the progressed Vertex is at 29:48 Virgo in the intercepted natal sign of Virgo and the progressed East Point is at 29:42 Pisces in the intercepted natal sign of Pisces. Progressed Mars at 25:52 Cancer is approaching the critical degree of 26 Cancer in the natal fifth house of romantic interests.

The progressed Moon at 10:10 Scorpio in the natal eighth house of shared resources, death and regeneration inconjuncts the progressed Sun at 11:20 Gemini in the natal third house of everyday environment. Progressed Mercury at 04:43 Cancer forms an exact conjunction with progressed Saturn at 04:43 Cancer in the natal fifth house of romantic interests. Mercury and Saturn square progressed Pluto at 04:06 Libra R and natal Pluto at 04:17 Libra R both in the seventh house of one-to-one relationships.

Thomas's Transits Interpreted for December 6, 1991

Transiting Pluto at the critical degree of 21:14 Scorpio transits the natal ninth house ruling foreign-born persons. Transiting Jupiter at 13:44 Virgo in the natal seventh house of one-to-one relationships forms an almost exact opposition to natal Jupiter at 13:45 Pisces in the natal first house of self. The exact opposition occurred on December 7, 1991. Both Virgo and Pisces are intercepted signs in the natal chart. Transiting Jupiter at 13:44 Virgo in the natal seventh house also separates from a square with the natal Midheaven at 13:24 Sagittarius. This square was exact on December 2, 1991.

Transiting Uranus at 12:15 Capricorn in the natal eleventh house of friends and acquaintances approaches a square with natal Venus at 12:47 Aries in the first house of self. This exact aspect occurred on December 16, 1991.

Lois Rodden, writer of *Modern Transits,* interprets these transiting aspects as follows:

> **Jup opp Jup (E)** -- Faith and trust can move mountains, but don't promise more than you can deliver.
> **Ura sqr Ven (A)** -- Unexpected changes in social life, and emotional relationships become unstable and volatile.

A Total Solar Eclipse on July 11, 1991 occurred at 18:59 Cancer. The eclipse falls in Thomas's fifth house and can indicate changes in school projects and romantic pursuits.

The Solar Return Chart for 1991 (Austin, TX)

Solar Uranus is at the critical degree of 13:30 Capricorn R in the tenth house of career and public recognition. It has no close aspects. Aries is intercepted in the solar first house of self and Libra is intercepted in the solar seventh house of one-to-one relationships. Mercury is placed at 29:19 in intercepted Aries in the first house of self, suggesting that Thomas's personality was more aggressive and impatient with others.

The solar Moon at 25:20 Gemini sextiles the Ascendant at 24:46 Pisces and opposes the Midheaven at 26:43 Sagittarius. Mars at 24:06 Cancer in the fifth house of romantic interests sextiles the Sun at 25:00 Taurus in the second house of values. Uranus also squares Fortuna or Misfortuna at 25:06 in the first house in the intercepted sign of Aries. Aries rules the head portion of the human body and also fire.

There is a yod in Thomas's solar return chart. It is comprised of the Sun at 25:00 Taurus in the second house of finances and values, Mars at 24:06 Cancer in the fifth house of romantic pursuits, and the Midheaven at 26:43 Sagittarius.

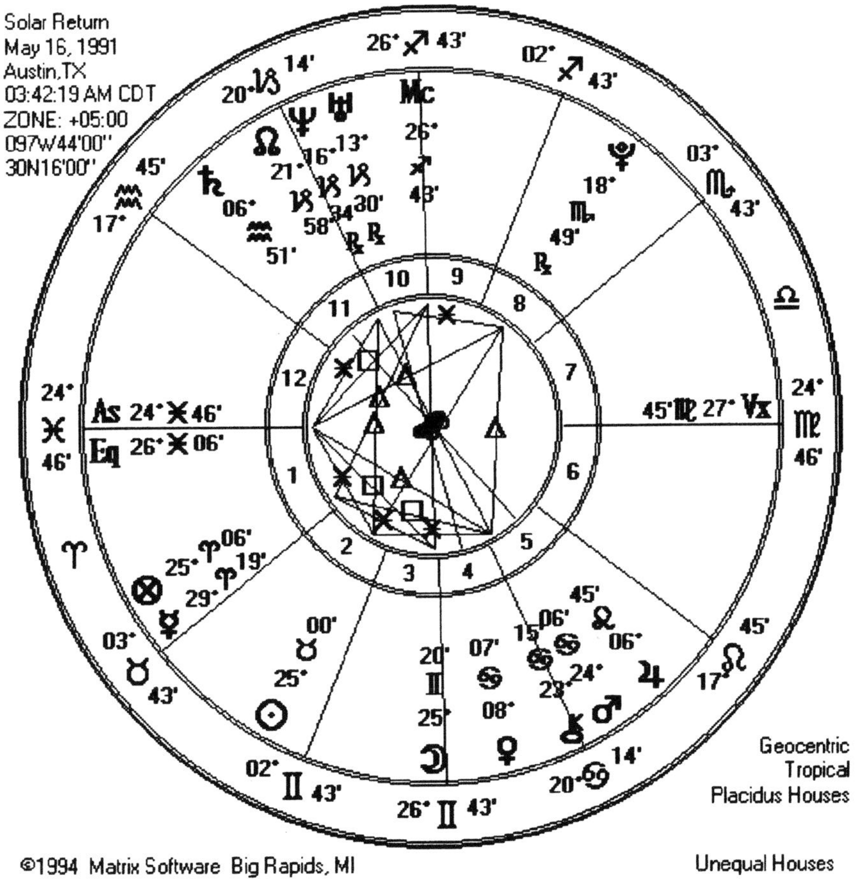

Chart 33 - Eliza Thomas's Solar Return Chart for 1991

The Yod or Fingers of Fate

There is one yod that should be noted in Thomas's combined charts. It is comprised of natal Sun at 25:00 Taurus, which is conjunct Algol, both in the natal third house of everyday environment. This conjunction sextiles progressed Mars at 25:52 Cancer in the natal fifth house of romantic relationships and inconjuncts transiting Moon at 25:17 Sagittarius in the tenth house of career environment and public recognition.

Following are the repetitive Basic Keywords from *The Rulership Book* by Rex E. Bills used to interpret the yod in Thomas's charts:

The Yod

Third House	Natal Sun	25:00 Taurus
	(The Natal Sun conjuncts Algol at 25:03 Taurus)	
Fifth House	Progressed Mars	25:52 Cancer
Tenth House	Transiting Moon	25:17 Sagittarius

The Yod includes the third, fifth and tenth houses

Third House, Natal 25:00 Taurus

Third House	**Sun**	**Taurus**
Acquaintances	Oneself	Base of the skull
Everyday environment	Men, in general in a woman's chart	Farms, farming, farmers, farm hands
Neighbors, neighborhood		

Fifth House, Progressed Mars 25:52 Cancer

Fifth House	**Mars**	
One's lovers, or rejected	Arson	Arguments
	Butchers	Assassinations, assassins
Cancer	Butchers	Carnage
Executions, executioners	Crimes of violence	Death, sudden
Milk, milkmen	Deaths by accident/violence	Fire, firearms
Cattle, cattle raisers cattlemen	Gunshot wounds	Gunman
	Head, diseases of head	Injuries by violence
	Males, in general	Murder
	Pimples on face, acne	Robbers, robbery
	Shooting, shootings	Slaughter, slaughter houses
	Men ages 25-35	Uncultured persons

Tenth House, Transiting Moon 25:17 Sagittarius

Tenth House	**Moon**	
Business, one's business life	Cows, cow barns	Crops, crop growers
	Deliveries, delivery men	Environment
Employers	Evening	Exhibitors, exhibitions
One's employment	Farmers, dairy	Gambling, gamblers
One's reputation	Milk, milkmen	Mental instabilities
Sagittarius		
Riding of horses	Dairies, dairy farmers	One's romantic tendencies
Stables	Places where dairy products are stored	Cattle, cattle raisers, cattlemen
Horses, horsemen		

The police theorized that the killer or killers involved in the Yogurt Shop Murders were transient to the Austin area and just passing through. However, the writer wrote a letter to a concerned party in Austin on June 2, 1992, four months before the issuance of the indictment for questioning of Carlos Saavedra and Alberto Cortes in October 1992. The letter listed the following items:

1. The girls may have know the killers, if only by sight.
2. The killers were from the general area.
3. Them killers have something to do with a "Farm."

4. Their general ages are 25-35.
5. The killers work around and ride horses.

The keywords picked out in Thomas's murder were the following:

Acquaintances | Everyday environment
Rejected lover? (No interest on Eliza's part, but the fifth house is activated by natal and progressed Mars, plus Solar Eclipse.)
Uncultured persons | Men, ages 25-35 **(The "Farm")**
Stock, stockyards, stock raisers | Cattle, cattle raisers, cattlemen
Cows, cow barns, cattlemen | Farms, farming, farmers, farm hands
(I Can't Believe It's Yogurt --
Is it made with Milk?) (What about Milk Delivery?)
Dairies, dairy farmers | Places where dairy products are stored
Delivery men, milk, milkmen

After the arrest of Carlos Saavedra and Alberto Cortes, the police released the following information about the men:

The three men lived in and had family in the general area of the yogurt shop.

The men's ages are: 22, 23 and 26.

Saavedra was released on parole from the Texas Department of Corrections in September 1991. A work "farm."

The three men kidnapped and sexually assaulted a woman in November 1991.

Saavedra may have gained access to the yogurt shop through his job as an ice cream delivery man.

In Conclusion

Eight years after the murders, Saavedra and Cortez are still in a Mexican prison serving time for cocaine possession and the 1991 rape of an Austin woman. The Mexican government does not extradite criminals. Therefore, the suspects have not been questioned since 1992 regarding their involvement in the Yogurt Shop Murder case. Much to the anguish of the families and friends of the victims, this crime remains unsolved.

The people of Austin have not forgotten that four of their teenagers died. Their security was shattered and many of their lives changed after that December night. Even today, there is sadness at the mention of these deaths. Many who were close to this case will never forget what they saw that night, and others remember the details of the girls' deaths. An article written by Janet Wilson entitled "Grief begins to lift for mother of 2 teens," tells of Barbara Suraci's coping with the death of her two daughters, Jennifer and Sarah Harbison. (See *Austin American-Statesman*, December 5, 1996.) Each of the girl's families have felt Suraci's heartbreak and their lives have been greatly affected by it.

In 1994, the families of the four girls settled a lawsuit with the owners of the yogurt shop and the shopping center for $12 million. The families published a manual in 1996, entitled "How to Survive and Thrive at Work: A Guide for Teenagers," which provides information they have learned since their daughter's deaths. They hope the information in this manual will save other children's lives. The parents also founded We Will Not Forget SAJE (Sarah, Amy, Jennifer and Eliza) which has a website dedicated to the girls at http://www.saje.org/home.htm.

Yogurt Shop Murders - October 1999 Update

In August 1999, the Austin Police Department assigned six investigators and a sergeant, and then enlisted the help of other agencies to pursue their latest lead. On October 6, 1999, police in West Virginia and Texas arrested Maurice Earl Pierce, 24, Michael James Scott, 25, Robert Burns Springsteen, Jr., 24, and Forrest Brook Welborn, 23, and charged them with capital murder in the deaths of Jennifer and Sarah Harbison, Amy Ayers and Eliza Thomas on December 6, 1991.

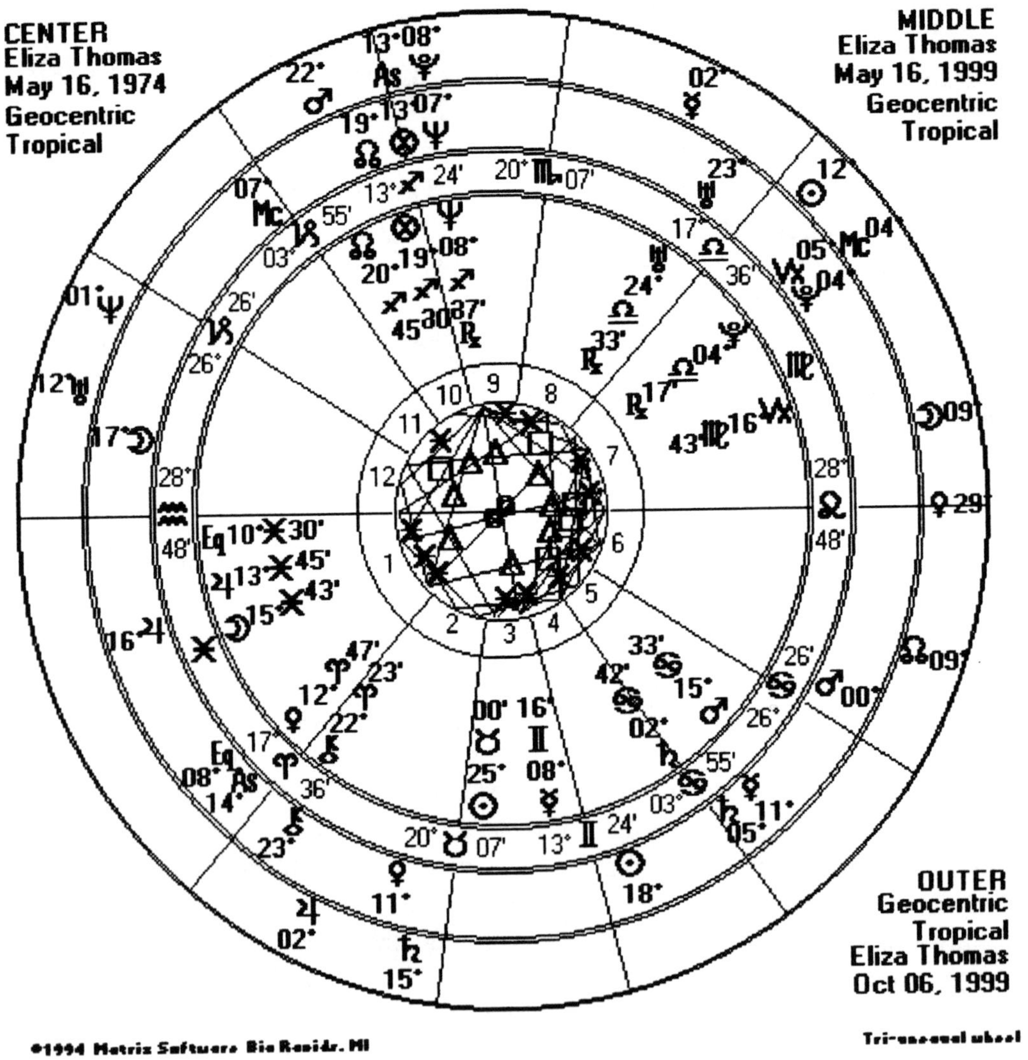

Extra Chart 33a - Eliza Thomas's Three-ring Chart with Natal, Progressed Chart for 1999 and Transits for Oct 6, 1999

Three weeks before the arrests, both Scott and Springsteen confessed to the crime and implicated Pierce and Welborn. The police statements say that Austin police questioned Pierce, 16, and Welborn, 15, days after the killings. At that time, Pierce told the officers that his gun had been used in the yogurt shop slayings. He further said that he had loaned his gun to a friend, Welborn, who had told him the details of the crime. Police questioned Welborn, but did not believe that he had committed the crime.

Sometime in 1996, Detective Paul Johnson of the Austin Police Department began reorganizing the massive case file of the murder case into manageable bits based on tips. Johnson later left the case and began working on it again when he was assigned to the Cold Cases Unit which is a group that investigates cases that have gone cold with time. According to reports, the Cold Cases Unit had 60 tips that it considered top priority. The re-examination led investigators back to the beginning – to that interview with an early suspect.

Maurice Pierce married a woman two years ago who he had dated for ten years and they have a seven-year old daughter. He was arrested at his home in Lewisville, about 25 miles northwest of Dallas. Records show that Pierce lived in Austin through November 1997. He was enrolled in different schools and attended Robbins Academy. That school is now closed, but served as a school for dropouts and students with discipline problems. Kimberli Pierce said that her husband has worked in warehouse jobs.

Forrest Welborn was the youngest of the four men arrested in connection with the case and is accused as being a lookout during the crime. Welborn lived in North Austin at the time and attended McCallum High School although he never graduated. His parents both work at the University of Texas. Welborn worked at Lockhart Muffler before it closed and then opened his own business, Lockhart Automotive, in April 1999. A co-worker said that he regarded Welborn as a talented mechanic and a good business partner.

Michael Scott began working at Collins Custom Care in early 1998. About the same time, he married Jeannine Stark who has a young daughter from a previous relationship. At the time of the murders, Scott was a sophomore at McCallum High School and attended special education classes. He dropped out of the Austin Schools about a year after the killings.

Robert Springsteen, Jr. attended alternative education classes for students with behavioral problems at McCallum High School. He dropped out about two weeks after the killings. He moved to West Virginia about eight years ago to live with his mother. Springsteen worked different jobs and was a manager at a MacDonald's restaurant until December 1998. He got married in February 1999.

Although Eliza Thomas had attended McCallum High School briefly before transferring to Lanier High School two months before the murders, it is not thought that she knew the four young men.

Birth Data for Eliza Hope Thomas:

Birth date:	May 16, 1974		
Birth time:	01:58 AM/EDT		
Birth place:	Boston, MA		
Longitude:	071W04	Latitude:	42N21
Source:	Birth certificate		

Birth Data for Other Girls without Birth Times:

Jennifer Harbison	Birth date:	May 5, 1974		
	Birth place:	Texarkana, TX		
	Longitude:	094W02	Latitude:	33N26
Sarah Harbison	Birth date:	October 28, 1976		
	Birth place:	Texarkana, TX		
	Longitude:	094W02	Latitude:	33N26
Amy Ayers	Birth date:	January 31, 1978		
	Birth place:	Cleburne, TX		
	Longitude:	097W23	Latitude:	32N21

Placements and degrees of Eliza Thomas' charts:
Natal Placements

Moon	15:43 Pis	01st	Neptune	08:37 Sag R	09th
Sun	25:00 Tau	03rd	Pluto	04:17 Lib R	07th
Mercury	08:16 Gem	03rd	North Node	20:45 Sag R	10th
Venus	12:47 Ari	01st	Vertex	16:43 Vir	07th
Mars	15:33 Can	05th	East Point	10:30 Pis	01st
Jupiter	13:45 Pis	01st	Fortuna	19:30 Sag	10th
Saturn	02:42 Can	04th	Midheaven	13:24 Sag	
Uranus	24:33 Lib R	08th	Ascendant	28:48 Aqu	

Eleventh house position is 03:55 Cap
Twelfth house position is 26:26 Cap
Second house position is 17:36 Ari
Third house position is 20:07 Tau

Critical Degrees: Venus approaching the critical degree of 13 Aries
To be noted: The Ascendant approaching the 29th degree of Aquarius
Intercepted Signs: Virgo intercepted in the seventh house of one-to-one relationships and Pisces intercepted in the first house of self

Progressed Placements for 1991

Moon	10:10 Sco	08th	Neptune	08:10 Sag R	09th

Sun	11:20 Gem	03rd	Pluto	04:06 Lib R	07th
Mercury	04:43 Can	05th	North Node	19:51 Sag R	10th
Venus	02:15 Tau	02nd	Vertex	29:48 Vir	07th
Mars	25:52 Can	05th	East Point	29:42 Pis	01st
Jupiter	15:55 Pis	01st	Fortuna	28:20 Leo	06th
Saturn	04:44 Can	05th	Midheaven	29:45 Sag	10th
Uranus	24:02 Lib R	08th	Ascendant	29:30 Pis	01st

Critical Degrees: Mars approaching the critical degree of 26 Cancer
To be noted: Vertex at 29 degrees of intercepted Virgo in the natal seventh house, East Point at the 29 degrees of intercepted Pisces in the natal first house, the Midheaven at 29 degrees of Sagittarius in the natal tenth house, and the Ascendant at 29 degrees of intercepted Pisces in the natal first house.

Transiting Placements for December 6, 1991 (10:15PM/EST)

Moon	25:02 Sag	10th	Uranus	12:15 Cap	11th
Sun	14:30 Sag	10th	Neptune	15:19 Cap	11th
Mercury	18:03 Sag R	10th	Pluto	21:14 Sco	09th
Venus	00:58 Sco	08th	North Node	11:07 Cap R	11th
Mars	05:44 Sag	09th	Vertex	26:19 Cap	11th
Jupiter	13:44 Vir	07th	East Point	20:41 Leo	06th
Saturn	03:18 Aqu	12th	Fortuna	12:03 Vir	07th

Critical Degrees: Pluto, Vertex
To be noted: Venus at 00 degrees of Scorpio

Solar Return Placements for 1991 (Austin, TX)

Moon	25:20 Gem	03rd	Neptune	16:34 Cap R	10th
Sun	25:00 Tau	02nd	Pluto	18:49 Sco R	08th
Mercury	29:19 Ari	01st	North Node	21:58 Cap R	11th
Venus	08:07 Can	04th	Vertex	27:45 Sco	07th
Mars	24:06 Can	05th	East Point	26:06 Pis	01st
Jupiter	06:45 Leo	05th	Fortuna	25:06 Ari	01st
Saturn	06:51 Aqu	11th	Midheaven	26:43 Sag	
Uranus	13:30 Cap R	10th	Ascendant	24:46 Pis	

Eleventh house position is 20:14 Cap
Twelfth house position is 17:45 Aqu

Second house position is 03:43 Tau
Third house position is 02:43 Gem

Critical Degrees: Uranus
To be noted: Mercury at 29 degrees of Aries in the first house of self. Aries also rules the head.

Intercepted Signs: Aries is intercepted in the first house of self and Libra is intercepted in the seventh house of one-to-one relationships.

Mary Fagan-Poleto
One of Albany's Unsolved Murders

Mary Fagan-Poleto, 30, grew up in the Johnstown and Perth area of upstate New York. She graduated from Perth High School in 1978 and then moved to Albany to attend college. She received an Associates Degree in Art from Maria College in 1982 and a Bachelor's Degree in Sociology from The College of Saint Rose in 1984 in Albany, NY. The longitude and latitude for Albany, NY are 73W45 and 42N39.

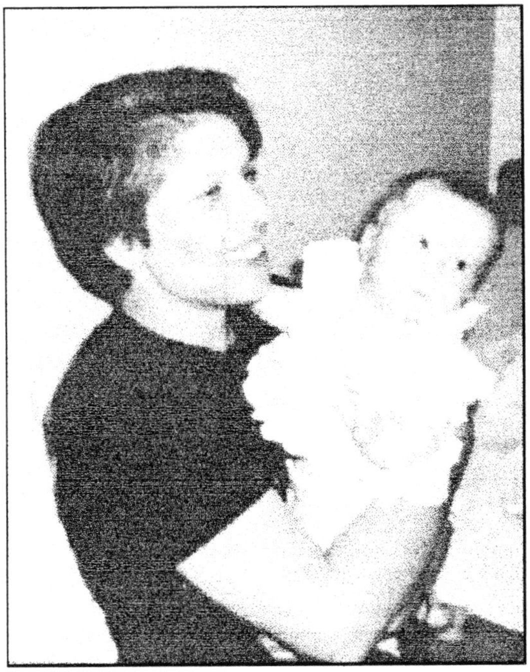

Picture 7 - Mary Fagan Poleto with Daughter

Fagan-Poleto lived a troubled life. She suffered from depression and was an out-patient of the Capital District Psychiatric Center in Albany where she received monthly treatments. These treatments made her drowsy and tired. At CDPC, she met another patient, James Poleto, and on June 17, 1989 they were married. Their daughter, Kyley Ann, was born in January 1991. Later, the couple separated and James had recently gained custody of their one-year-old daughter. Fagan-Poleto was considering reconciling with her husband so that she could be with her daughter.

On September 4, 1991, Fagan-Poleto received her monthly treatment at CDPC. As usual she was in a state of drowsiness when she went to bed. About 1:00 AM/EST on September 5, someone entered her bedroom while she slept and stabbed her in the back. Unable to move, Fagan-Poleto could only lay there and bleed to death of the single stab wound that had injured her spinal column.

Fagan-Poleto had not been sexually assaulted. The police thought Mary had let her killer in even though a screen had been removed from her second-floor bedroom window. The Albany police did have several clues in Fagan-Poleto's murder, but they did not have a suspect.

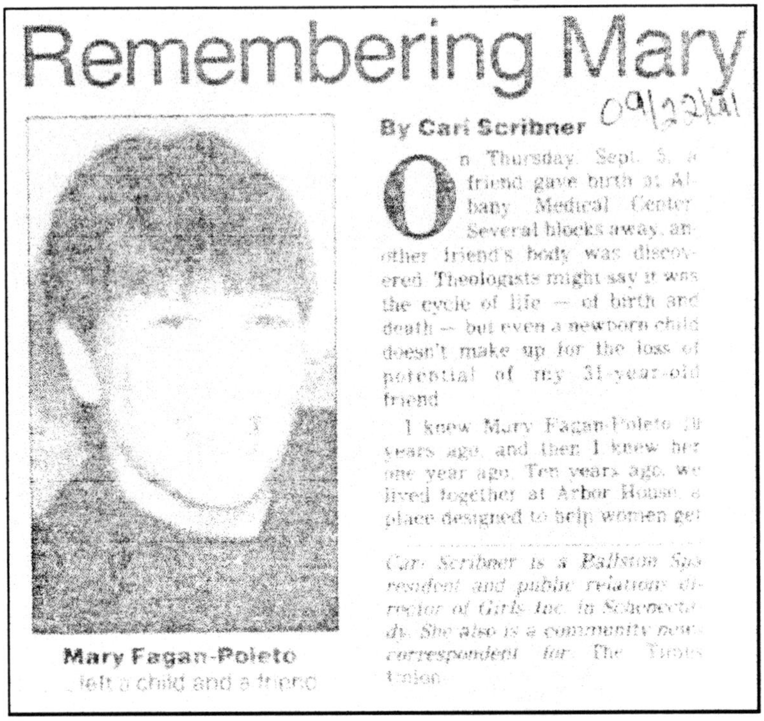

Remembering Mary

09/22/01

By Cari Scribner

On Thursday, Sept. 5, a friend gave birth at Albany Medical Center. Several blocks away, another friend's body was discovered. Theologists might say it was the cycle of life — of birth and death — but even a newborn child doesn't make up for the loss of potential of my 31-year-old friend.

I knew Mary Fagan-Poleto 10 years ago, and then I knew her one year ago. Ten years ago, we lived together at Arbor House, a place designed to help women get

Cari Scribner is a Ballston Spa resident and public relations director of Girls Inc. in Schenectady. She also is a community news correspondent for The Times Union.

Mary Fagan-Poleto
left a child and a friend

Illustration 21 - Albany Times Union (Fagan-Poleto)

Fagan-Poleto's Natal Chart

Natal Neptune is at the critical degree of 09:55 Scorpio in the third house of everyday environment. It sextiles natal Pluto at 08:06 in the first house of self. Natal Sun at 07:56 Sagittarius in the fourth house of home and personal matters also squares natal Pluto. Mercury at 19:14 Scorpio in the third house trines Mars at 18:02 Cancer R in the eleventh house of friends and acquaintances and opposes the Midheaven at 20:01 Taurus.

Natal Venus at 17:49 Capricorn in the fifth house of romantic relationships conjuncts Saturn at 16:03 Capricorn. Venus and Saturn oppose Mars at 18:02 Cancer R in the eleventh house. Jupiter at 06:53 Capricorn in the fifth house trines Pluto at 08:06 Virgo in the first house of self.

The Progressed Chart for 1990

Progressed Jupiter is at the critical degree of 13:40 Capricorn in the natal fifth house of children and romantic relationships. The progressed Moon at 07:23 Gemini in the natal tenth house of career and public recognition inconjuncts the progressed Sun at 08:27 Capricorn in the natal fifth house and squares natal Pluto at 08:06 Virgo R and progressed Pluto at 08:02 Virgo R both in the natal first house of self. The progressed

Sun also opposes progressed Mars at 08:51 Cancer R in the natal eleventh house of friends and acquaintances.

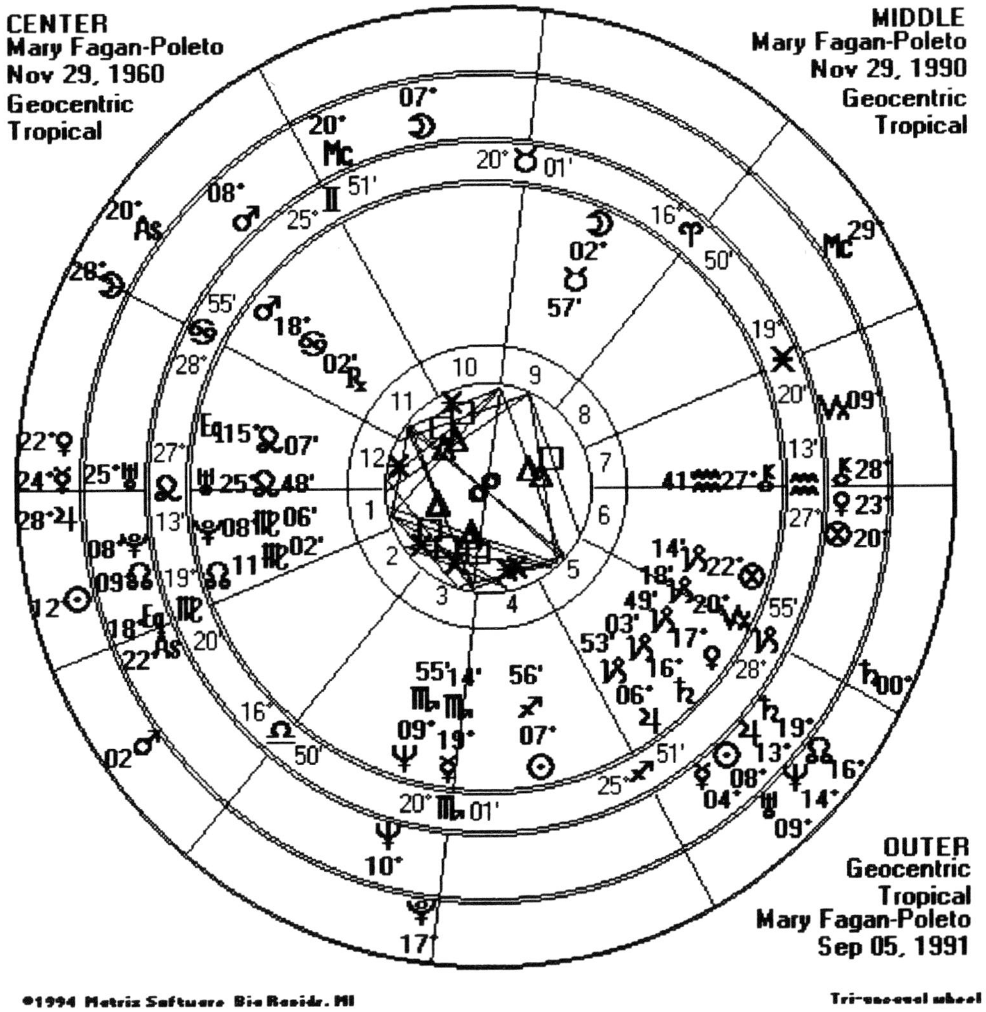

Chart 34 - Mary Fagan Poleto's Three-ring Chart with Natal, Progressed Chart for 1990 and Transits for September 5, 1991

Progressed Venus at 23:10 Aquarius in the natal sixth house of health and service to others opposes progressed Uranus at 25:26 Leo R in the natal twelfth house of institutions, self-undoing and secret enemies. Progressed Venus also inconjuncts the progressed Ascendant at 22:00 Virgo. Progressed Saturn at 19:22 Capricorn in the natal fifth house inconjuncts the progressed Midheaven at 20:32 Gemini. Progressed Neptune at 10:48 Scorpio in the natal third house of everyday environment sextiles the progressed Sun at 08:27 in the natal fifth house and progressed Pluto at 08:02 Virgo R and the progressed North Node at 09:26 Virgo R, both in the first house.

Progressed Fortuna or Misfortuna at 20:56 Aquarius is approaching the critical 21 degrees of Aquarius in the natal seventh house of one-to-one relationships. It trines the progressed Midheaven at 20:32 Gemini and inconjuncts the progressed Ascendant at 22:00 Virgo.

Fagan-Poleto's Transits Interpreted for September 5, 1991

Transiting Pluto at 17:58 in the natal third house of everyday environment sextiles natal Venus at 17:49 Capricorn in the natal fifth house of children and love relationships. Transiting Uranus at 09:55 Capricorn R in the natal fifth house forms an exact sextile with natal Neptune at 09:55 Scorpio in the third house of everyday surroundings. Look at this exact Uranus / Neptune aspect closer:

Uranus is the planet of the sudden and unexpected
Capricorn, a feminine sign, rules the older or authoritative person
Fifth house rules romantic relationships

Neptune is the planet of the secret and hidden
Scorpio, a feminine sign, rules sex and death
Third house rules everyday environment

Transiting Pluto at 17:58 Scorpio in the natal third house trines natal Mars at 18:02 Cancer R in the eleventh house of friends and acquaintances. This aspect is exact on September 7, 1997.

Lois Rodden, writer of *Modern Transits,* interprets these transiting aspects as follows:

Pluto sxt Venus (S) -- In your social life there are opportunities to meet a variety of people that can bring you a deep friendship or emotional impact.

Uranus sxt Neptune (E) -- Independent aspirations bring opportune advantages; however, erratic discontent opens strange doors.

Pluto trine Mars (A) -- Under stress, competition and repression you can transmute your passion into constructive action.

A Total Solar Eclipse occurred on July 11, 1991 at 18:59 Cancer. The eclipse falls in Fagan-Poleto's eleventh house and can indicates changes in friends, acquaintances and long-term goals.

The Solar Return Chart for 1990 (Albany, NY)

Solar Mars is at the critical degree of 04:23 Gemini R in the seventh house of one-to-one relationships. Mars forms no close aspects, but opposes the solar Sun at 07:56 in the first house of self. The solar Moon at 10:50 Taurus in the solar sixth house of health and service trines Neptune at 12:59 Capricorn, which is approaching the critical 13 degrees in the second house of finances and values.

Solar Venus at 15:05 Sagittarius in the second house trines Jupiter at 13:36 Leo R in the ninth house of abstract mind and Fortuna or Misfortuna at 16:27 Aries in the fifth house of children and romantic affairs. Solar Jupiter at 13:36 Leo R forms a square with the Solar Ascendant at 13:34 Scorpio.

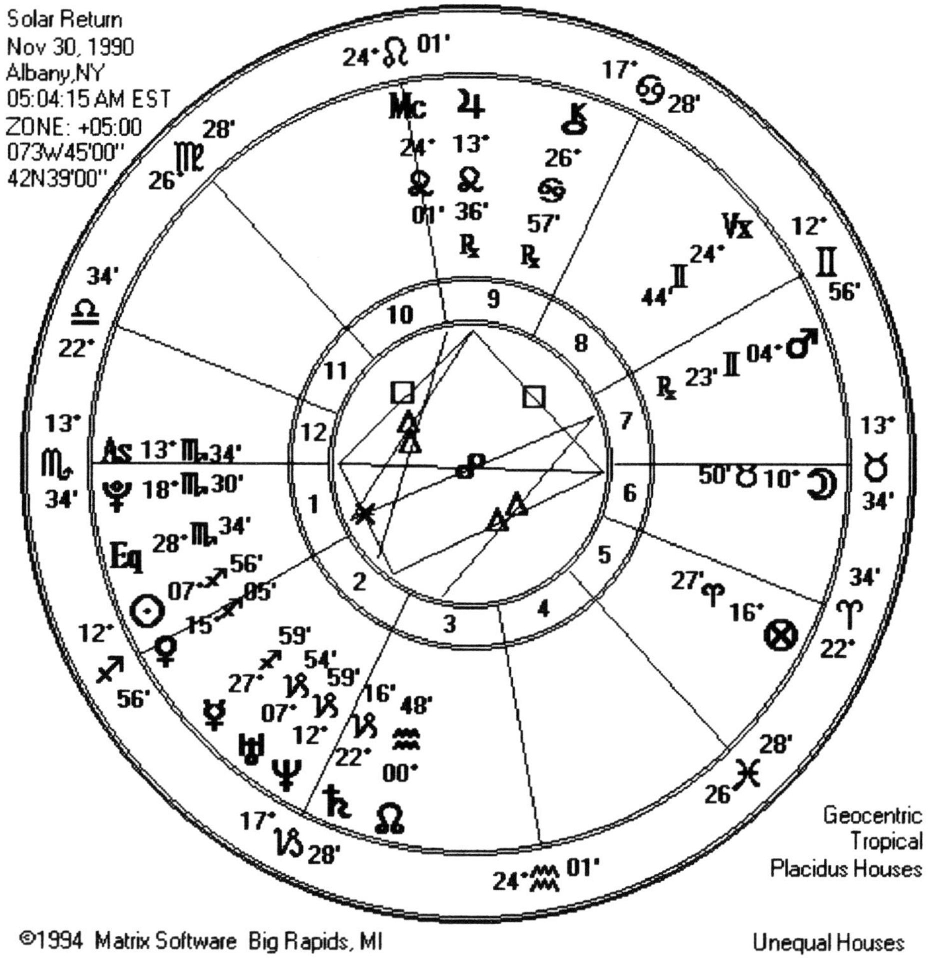

Chart 35 - Mary Fagan-Poleto's Solar Return Chart for 1990

The Yod or Fingers of Fate

There is one yod that should be noted in Fagan-Poleto's combined charts. It is comprised of natal Neptune at 09:55 Scorpio and progressed Neptune at 10:48 Scorpio both in the natal third house of everyday environment, sextile natal Jupiter at 06:53 Capricorn, progressed Sun at 08:27 Capricorn and transiting Uranus at the exact degree of natal Neptune of 09:55 Scorpio R all in the natal fifth house of love relationships. These planets for an inconjunct to the progressed Moon at 07:23 Gemini in the natal tenth house of public recognition.

In Conclusion

This is no suspect in Mary Fagan-Poleto's murder and the investigation into her death remains open.

Birth Data for Mary Fagan-Poleto

Birth date:	November 29, 1960		
Birth time:	10:31 PM/EST		
Birth place:	Amsterdam, NY		
Longitude:	074W11	Latitude: 42N56	
Source:	From family		

Daughter:

Kyley Ann:	Birth date:	January 28, 1991	
	Birth time:	02:33 PM/EST	
	Birth place:	Albany, NY	
	Longitude:	073W47	Latitude: 42N39
	Source:	From family	

Husband:

James:	Birth date:	November 1, 1957	
	Birth time:	09:47 AM/EST	
	Birth place:	Albany, NY	
	Longitude:	073W47	Latitude: 42N39
	Source:	From family	

Placements and degrees of Mary's charts:
Natal Placements

Moon	02:57 Tau	09th	Neptune	09:55 Sco	03rd
Sun	07:56 Cap	04th	Pluto	08:06 Vir	01st
Mercury	19:14 Sco	03rd	North Node	11:02 Vir R	01st
Venus	17:49 Cap	05th	Vertex	20:18 Cap	05th
Mars	18:02 Can R	11th	East Point	15:07 Leo	12th
Jupiter	06:53 Cap	05th	Fortuna	22:14 Cap	05th
Saturn	16:03 Cap	05th	Midheaven	20:01 Tau	
Uranus	25:48 Leo	12th	Ascendant	27:13 Leo	

Eleventh house position is 25:51 Gem
Twelfth house position is 28"55 Can

Second house position is 19:20 Vir
Third house position is 16:50 Lib

Critical Degrees: Neptune
To be noted: None
Intercepted Signs: None

Progressed Placements for 1990

Moon	07:23 Gem	10th	Neptune	10:48 Sco	03rd
Sun	08:27 Cap	05th	Pluto	08:02 Vir R	01st

Mercury	04:29 Cap	05th	North Node	09:27 Vir R	01st
Venus	23:10 Aqu	06th	Vertex	09:23 Pis	07th
Mars	08:51 Can R	11th	East Point	18:48 Vir	01st
Jupiter	13:40 Cap	05th	Fortuna	20:56 Aqu	06th
Saturn	19:22 Cap	05th	Midheaven	20:32 Gem	10th
Uranus	25:26 Leo R	12th	Ascendant	22:00 Vir	02nd

Critical Degrees: Jupiter
To be noted: Fortuna or Misfortuna approaching 21 degrees of Aquarius

Transiting Placements for September 5, 1991 (1:00 AM/EST)

Moon	28:39 Can	11th	Uranus	09:55 Cap R	05th
Sun	12:12 Vir	01st	Neptune	14:06 Cap R	05th
Mercury	24:34 Leo	12th	Pluto	17:58 Sco	03rd
Venus	22:19 Leo R	12th	North Node	16:02 Cap R	05th
Mars	02:34 Lib	02nd	Vertex	06:35 Sag	04th
Jupiter	28:30 Leo	01st	East Point	29:42 Gem	11th
Saturn	00:54 Aqu R	06th	Fortuna	06:30 Gem	10th

Critical Degrees: None
To be noted: Saturn at 00 degrees of Aquarius and the East Point at 29 degrees of Gemini

Solar Return Placements for 1990 (Albany, NY)

Moon	10:50 Tau	06th	Neptune	12:59 Cap	02nd
Sun	07:56 Sag	01st	Pluto	18:30 Sco	01st
Mercury	27:59 Sag	02nd	North Node	00:48 Aqu R	03rd
Venus	15:05 Sag	02nd	Vertex	24:44 Gem	08th
Mars	04:23 Gem R	07th	East Point	28:34 Sco	01st
Jupiter	13:36 Leo R	09th	Fortuna	16:27 Ari	05th
Saturn	22:16 Cap	03rd	Midheaven	24:01 Leo	
Uranus	07:54 Cap	02nd	Ascendant	13:34 Sco	

Eleventh house position is 26:29 Vir
Twelfth house position is 22:35 Lib
Second house position is 12:56 Sag
Third house position is 17:28 Cap

Critical Degrees: Mars
To be noted: Neptune approaching the critical degree of 13 Capricorn and the North Node at 00 Aquarius
Intercepted Signs: None

Daniel Vallone
The Cabbagetown "Mayor"

Daniel Vallone, 36, won his title as the unofficial mayor of Cabbagetown with his protester's heart and hard work on restoration projects in the southeast Atlanta, GA neighborhood. He had also just completed the restoration of his home and was planning a celebration. Vallone made plans to have the party on December 4, 1987. In the midst of these events, he thought he had inadvertently misplaced his checkbook. After contacting the bank, he found that it had been stolen and that someone had forged several checks. He recognized the signature on the checks as that of James Malloy.

Malloy, a transient, was a member of an area foundation crew who later befriended Vallone and stayed on as a carpenter to help him with the reconstruction of his house. However, what Vallone did not know was that Malloy was wanted in six Georgia counties for several different felony charges.

Picture 8 - Daniel Vallone

On December 3, 1987, Vallone disappeared from his home in the suburbs of Atlanta. On that night, he had spoken with several people regarding the party the following night. However, in the morning Vallone and his late model pickup truck were missing. Ironically, Malloy had also checked out of his motel room early on the morning of December 4.

During the next few weeks, Vallone's credit cards were used to charge Malloy's expenses from Georgia to Michigan. Vallone's truck was found in Gasney, SC on February 11, 1988 burned almost beyond recognition. On October 9, 1988, this case was featured on the Fox Network's program "America's Most Wanted." As a result of this show, Malloy was captured in Tampa, FL on November 17, 1988.

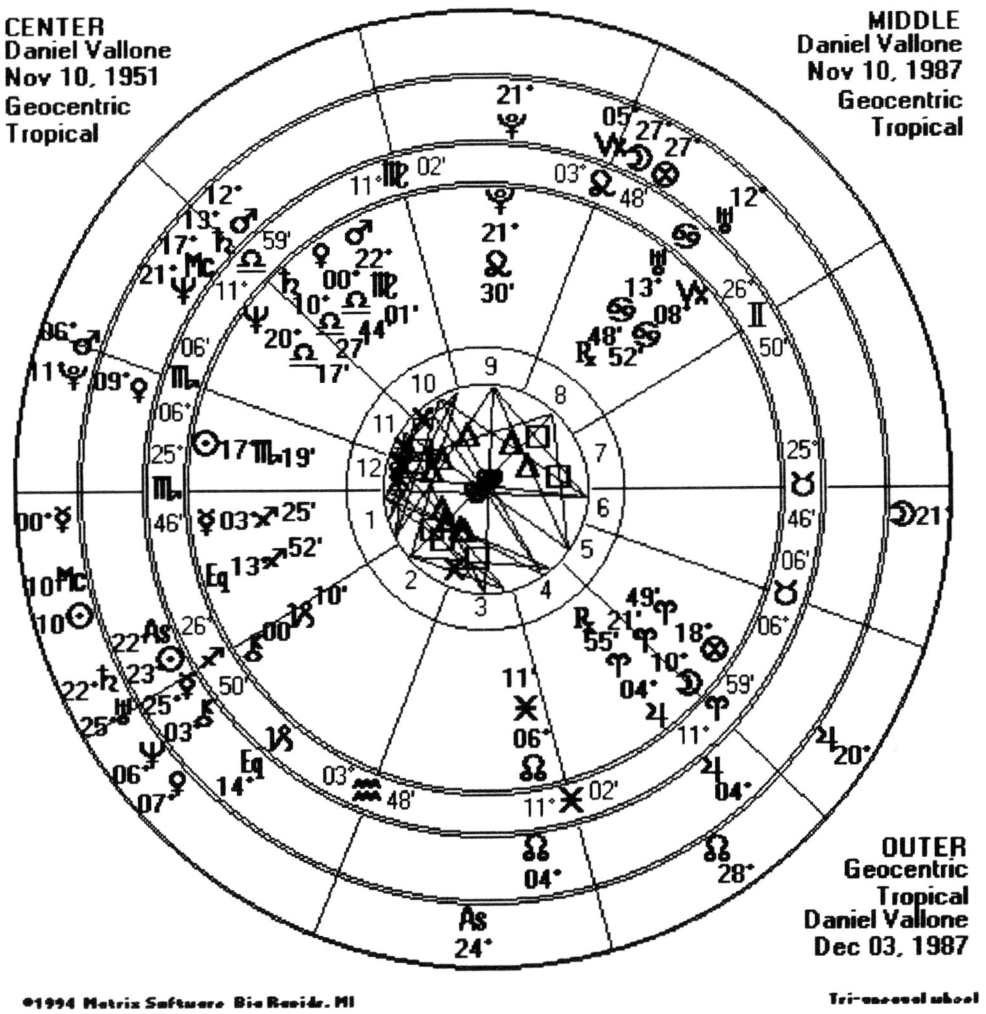

Chart 36 - Daniel Vallone's Three-ring Chart with Natal, Progressed Chart for 1987 and Transits for December 3, 1987

Natal Charts

Using the birth charts of both men, it is interesting to see how they interacted. Astrology points out who was the victim and who was the assailant. The following are interpretations in part from *Predictive Astrology* by Frances Sakoian and Louis S. Acker of Vallone's and Malloy's astrological transits or trends for December 3, 1987. This was the last day Vallone was seen and last spoken to at between 7:30 and 8:00 P.M.

Vallone: On the night of December 3, Vallone was prone to deception through others in financial affairs and would push against a distinct obstacle. (Malloy) (Moon inc Neptune - A) ** Loss or theft of personal property and valuables through dealing with a criminal element. (Vallone knew Malloy had stolen his checkbook and had written checks against the account.) (Venus ses Pluto - S 00:29) ** Disagreements over finances possibly because of extortion or coercion. (Vallone's confrontation with Malloy over the forged checks.) (Venus ses Pluto - S 00:29) ** Caution should be used in all financial dealings. (Venus ses Pluto - S 00:29) ** Financial loss through theft. (Venus ses Pluto - S 00:29) ** There was danger from confidence artists. (Jupiter opp Neptune - S 00:17) ** Exposure to illegal acts. Loss through extortioners, swindlers and confidence men. (Jupiter opp Neptune - S 00:17)

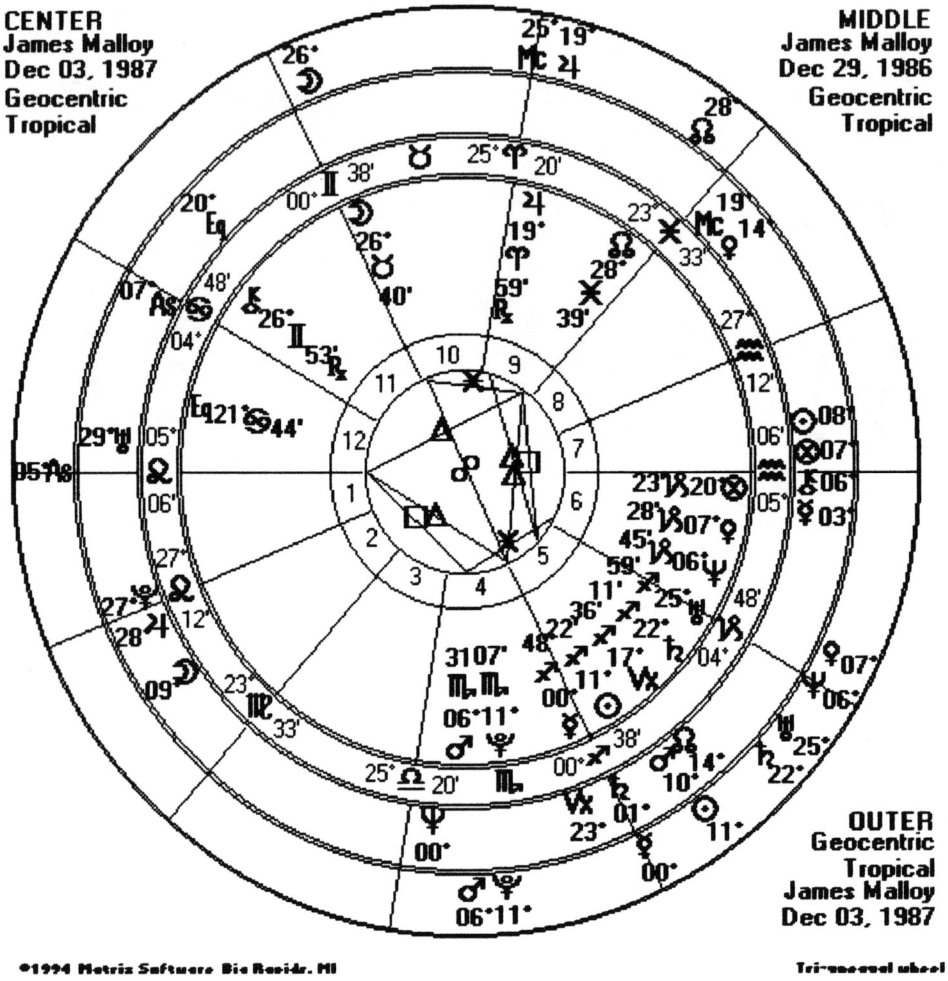

Chart 37 - James Malloy's Three-ring Chart with Natal, Progressed Chart for 1986 and Transits for December 3, 1987

This was a time for Vallone to do physical work involving the changing or moving of his residence and/or a time for doing construction or renovations on his home. (This was evidenced by the completion of the restoration of his home.) (Moon tri Mars - A 00:13)

Further aspects read: A danger in dealing with the criminal element. (Malloy) (Venus ses Pluto - S 00:29) ** Someone in Vallone's environment was obnoxious, hostile, or competitive. (Malloy) (Mars ssq Mars A 00:45) ** Disputes, violent acts, tendency to suffer injuries, destruction. There was a need for caution in dealing with fire arms, explosives, and dangerous materials. (Mars ssq Mars A 00:45) ** Impulsive, aggressive behavior characterized this period. (Mars ssq Mars A 00:45) ** Anger, fights, battles, outbursts of temper and, in extreme cases, physical confrontations were possible.

Accidents or injuries may result. (Did Vallone confront Malloy over the missing checkbook and forged checks? Did a fight between the two develop?) This was not a time for Vallone to meet head on with another. (Mars ssq Mars A 00:45)

Malloy: A look at Malloy's chart for the same day may possibly tell the other half of this disappearance and homicide:

Malloy was experiencing a sense of inferiority, a lack of confidence and a fear of exposure. (Moon opp Saturn A 02:01) ** His lowered vitality and depression characterized this period. (Moon opp Saturn A 02:01) ** This was not a time for Malloy to deal with women or older people. (Moon opp Saturn A 02:01) ** He was prone to sudden outbursts of emotion, confrontations with subconscious problems, extreme emotional reactions, and the eruption of deep-seated emotional problems. (Moon sqq Jupiter A 01:45) ** He had a short fuse on anger and impatience that may hamper his judgment. ** There was tactlessness and lack of self-control on his part. (Mars sqq Venus A 00:12) ** Jealousy. (Mars sqq Venus A 00:12) ** He was unreliable, dishonest and had a speculative failure. (Jupiter sqq Mercury S 01:20) ** He suffered from over-expanded optimism. (Jupiter sqq Mercury S 01:20)

Malloy had poor judgment and was prone to engage in unethical business profiteering. (Jupiter inc Mars A 00:02) ** He had an inflated sense of one's own importance and a tendency to take too much for granted. (Jupiter inc Mars A 00:02) ** There was a lack of moderation. (Jupiter inc Mars A 00:02) ** He would take unnecessary financial and physical risks. (Jupiter inc Mars A 00:02) ** A time of negative character, tendency to alcohol, confusion, conflicts arising from jealousy and immoral acts. (Neptune con Sun A 00:27) ** There would be problems with joint finances and goods of the dead. (Jupiter inc Mars A 00:02) ** Minor changes in his work may have caused an emotional upset. (The house had just been completed.) (Saturn ssq Venus S 00:28) ** Avoid getting into a critical fight, conflicts, or performing unlawful acts. (Mercury sqq Jupiter A 00:29) ** Unexpected travel. (Mars trine Uranus A 00:26)

A Annular Solar Eclipse on September 23, 1987 occurred at the decisive degree of 29:33 Virgo. The eclipse falls in Vallone's eleventh house and can indicate changes in friends, acquaintances and long-term goals. The same Annular Solar Eclipse falls in Malloy's fifth house and can indicate changes in love relationships, children and creativity.

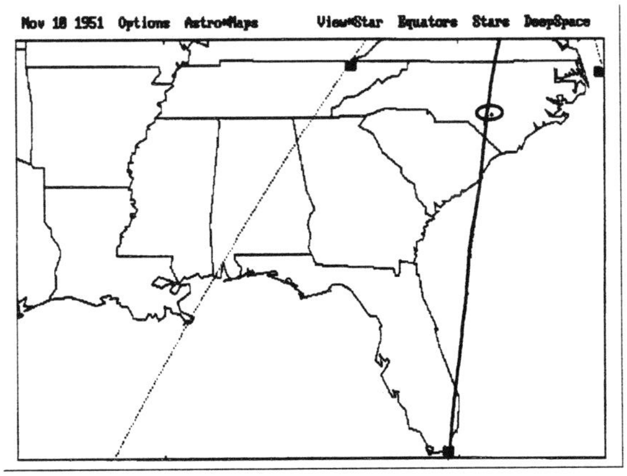

Map 7 - Daniel Vallone's Natal Local Space Map

Astro*Carto*Graphy

There are some interesting factors in the Astro*Carto*Graphy maps for Vallone and Malloy. In Vallone's natal local space map the North Node runs close to Fayetteville, NC. Malloy was born in Fayetteville. In Vallone's transit map for December 3, 1987 Mercury the ruler of friends runs through Fayetteville.

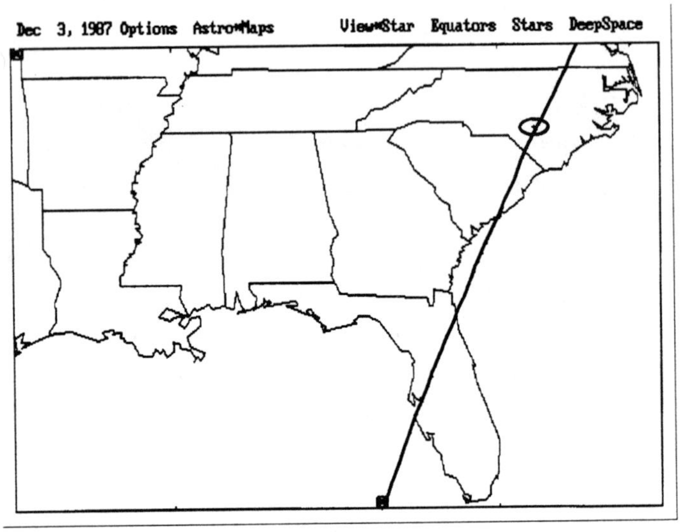

Map 8 - Daniel Vallone's Transit Cyclo*Carto*Graphy Map for December 3, 1987

In Conclusion

In January 1996, a man found a human skull while walking his dog on his farm in Newton County, GA about 40 miles east of Atlanta. Identification of the remains ended the nine-year mystery of what happened to Daniel Vallone. However, it is still unclear to officials who had crushed his head with a blunt object and left him for dead.

Birth Data for Daniel Vallone

Birth date:	November 10, 1951
Birth time:	07:45 AM/EST
Birth place:	Rochester, NY
Longitude:	077W37 Latitude: 43N09
Source:	From mother

Birth Data for James Malloy

Birth date:	December 29, 1955
Birth time:	03:06 PM/EST
Birth place:	Fayetteville, NC
Longitude:	078W53 Latitude: 35N03
Source:	Birth certificate

Placements and degrees of Vallone's charts:
Natal Placements

Moon	10:21 Ari	04th	Neptune	20:17 Lib	11th
Sun	17:19 Sco	12th	Pluto	21:30 Leo	09th
Mercury	03:25 Sag	01st	North Node	06:11 Pis R	03rd
Venus	00:44 Lib	10th	Vertex	08:52 Can	08th

Mars	22:01 Vir	10th	East Point	13:52 Sag	01st
Jupiter	04:55 Ari R	04th	Fortuna	18:49 Ari	05th
Saturn	10:27 Lib	10th	Midheaven	11:02 Vir	
Uranus	13:48 Can R	08th	Ascendant	25:46 Can	

Eleventh house position is 11:59 Lib
Twelfth house position is 06:06 Sco

Second house position is 26:50
Third house position is 03:48 Aqu

Critical Degrees: Venus, Uranus, Pluto, and Ascendant approaching 26:00 Cancer
To be noted: None
Intercepted Signs: Cancer in the eighth house and Capricorn in the second house

Progressed Placements for 1987

Moon	27:53 Can	08th	Neptune	21:18 Lib	11th
Sun	23:45 Sag	01st	Pluto	21:26 Leo R	09th
Mercury	25:12 Sag R	01st	North Node	04:17 Pis R	03rd
Venus	09:44 Sco	12th	Vertex	05:15 Leo	09th
Mars	12:28 Lib	11th	East Point	14:51 Cap	02nd
Jupiter	04:42 Ari	04th	Fortuna	27:06 Can	08th
Saturn	13:38 Lib	11th	Midheaven	17:29 Lib	11th
Uranus	12:42 Can R	08th	Ascendant	22:58 Sag	01st

Critical Degrees: Venus, Saturn, Pluto, and North Node
To be noted: None

Transiting Placements for December 3, 1987

Moon	21:49 Tau	06th	Uranus	25:58 Sag	01st
Sun	10:59 Sag	01st	Neptune	06:44 Cap	02nd
Mercury	00:13 Sag	01st	Pluto	11:07 Sco	12th
Venus	07:00 Cap	02nd	North Node	28:40 Pis R	04th
Mars	06:16 Sco	11th	East Point	07:42 Pis	03rd
Jupiter	20:00 Ari R	05th	Vertex	14:39 Vir	10th
Saturn	22:08 Sag	01st	Fortuna	04:52 Leo	09th

Critical Degrees: Moon
To be noted: Mercury at 00 degrees
Solar Return Placements for 1987 (Atlanta, GA)

Moon	10:50 Can	11th	Neptune	06:00 Cap	05th
Sun	17:19 Sco	03rd	Pluto	10:12 Sco	03rd
Mercury	28:45 Lib	02nd	North Node	29:54 Pis R	07th
Venus	07:51 Sag	04th	Vertex	22:51 Cap	05th
Mars	20:58 Lib	02nd	East Point	28:37 Leo	12th

Jupiter	21:49 Ari R	08th	Fortuna	28:55 Ari	08th
Saturn	19:30 Sag	04th	Midheaven	02:50 Gem	
Uranus	24:39 Sag	04th	Ascendant	05:24 Vir	

Eleventh house position is 05:50 Can
Twelfth house position is 07:00 Leo
Second house position is 00:42 Lib
Third house position is 00:15 Sco

Critical Degrees: None
To be noted: North Node at 29 degrees of Pisces and Fortuna or Misfortuna approaching 29 degrees of Aries
Intercepted Signs: None

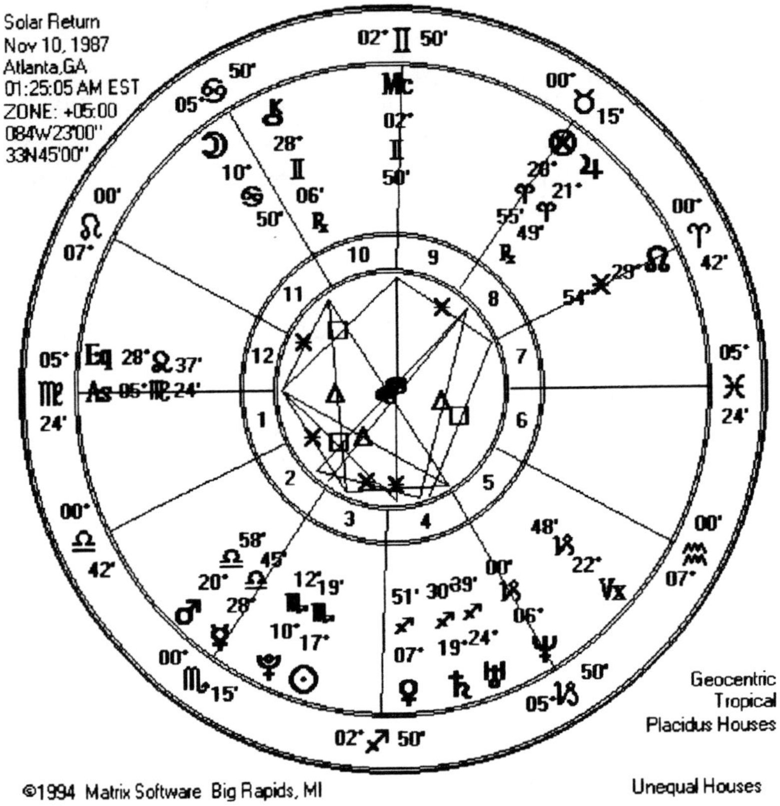

Chart 38 - Daniel Vallone's Solar Return Chart for 1987

Placements and degrees of James Malloy's charts:
Natal Placements

Moon	17:03 Can	02nd	Neptune	06:00 Sco	06th
Sun	07:20 Cap	08th	Pluto	28:25 Leo R	04th
Mercury	21:31 Cap	08th	North Node	16:12 Sag R	07th
Venus	06:53 Aqu	09th	Vertex	03:35 Sco	06th
Mars	20:01 Sco	06th	East Point	22:32 Tau	12th
Jupiter	01:17 Vir R	04th	Fortuna	17:46 Sag	07th

Saturn	28:40 Sco	06th	Midheaven	17:40 Aqu	
Uranus	01:14 Leo R	03rd	Ascendant	08:03 Gem	

Eleventh house position is 18:03 Pis
Twelfth house position is 27:00 Ari
Second house position is 01:46 Can
Third house position is 23:34 Can

Critical Degrees: Fortuna
To be noted: None
Intercepted Signs: Taurus in the twelfth house of confinement and self-undoing and Scorpio in the sixth house of service and occupation

Progressed Placements for 1986

Moon	09:36 Vir	04th	Neptune	00:25 Sco	06th
Sun	08:55 Aqu	09th	Pluto	27:49 Leo R	04th
Mercury	03:48 Aqu R	09th	North Node	14:34 Sag R	07th
Venus	14:56 Pis	10th	Vertex	23:40 Sco	06th
Mars	10:19 Sag	07th	East Point	20:55 Gem	01st
Jupiter	28:45 Leo R	04th	Fortuna	07:43 Aqu	09th
Saturn	01:20 Sag	06th	Midheaven	19:15 Aqu	
Uranus	29:55 Can R	03rd	Ascendant	07:02 Gem	

Critical Degrees: Sun approaching 09 degrees of Aquarius
To be noted: Jupiter approaching 29 degrees of Leo, Uranus at 29 degrees of Cancer and Neptune at 00 degrees of Scorpio

Transiting Placements for December 3, 1987 (9:00 P.M.)

Moon	26:40 Tau	12th	Uranus	25:59 Sag	07th
Sun	11:22 Sag	07th	Neptune	06:45 Cap	08th
Mercury	00:48 Sag	06th	Pluto	11:07 Sco	06th
Venus	07:28 Cap	08th	North Node	28:39 Pis R	11th
Mars	06:31 Sco	06th	Vertex	17:36 Sag	07th
Jupiter	19:59 Ari R	11th	East Point	21:44 Can	02nd
Saturn	22:11 Sag	07th	Fortuna	20:23 Cap	08th

Critical Degrees: None
To be noted: Mercury at 00 degrees of Sagittarius in the sixth house of service to others

Solar Return Placements for 1986 (Atlanta, GA)

Moon	12:12Sag	01st	Neptune	05:35 Cap	02nd
Sun	07:20 Cap	02nd	Pluto	09:24 Sco	12th
Mercury	29:05 Sag	02nd	North Node	16:38 Ari R	05th
Venus	21:48 Sco	01st	Vertex	06:13 Can	08th
Mars	22:55 Pis	04th	East Point	28:53 Sco	01st
Jupiter	17:06 Pis	04th	Fortuna	22:11 Lib	11th

Saturn	15:05 Sag	01st	Midheaven	24:21 Leo
Uranus	23:27 Sag	02nd	Ascendant	17:19 Sco

Eleventh house position is 26:40 Vir
Twelfth house position is 24:13 Lib
Second house position is 16:54 Sag
Third house position is 19:49 Cap

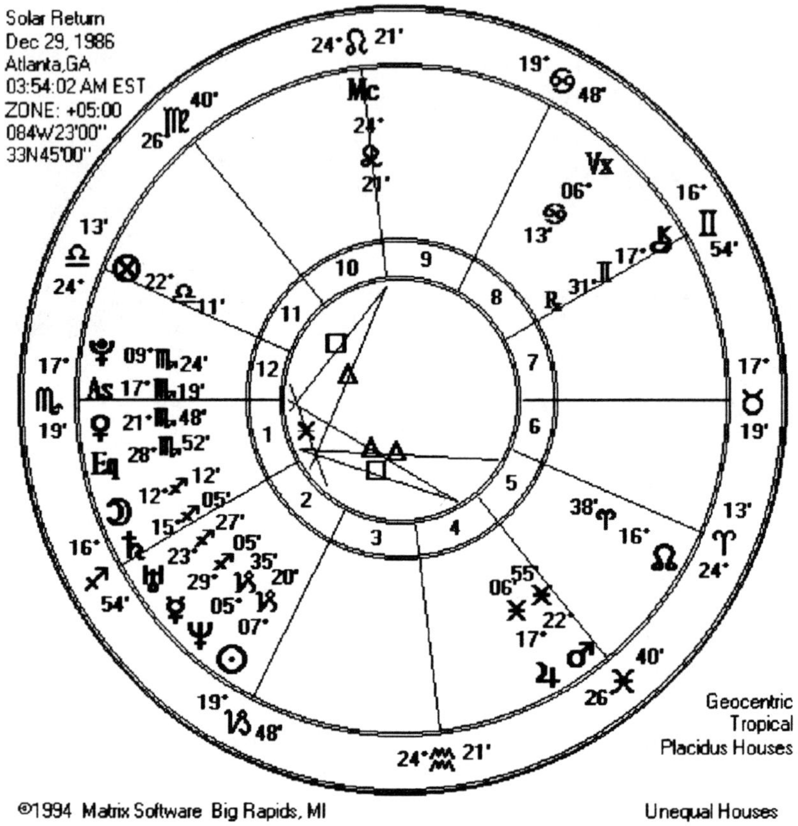

Chart 39 - James Malloy's Solar Return Chart for 1986

Critical Degrees: Venus, Jupiter, and Pluto
To be noted: Mercury at 29 degrees of Sagittarius, and East Point approaching 29 degrees of Scorpio
Intercepted Signs: None

MURDER, ACCIDENT OR SUICIDE?

Eric Paquette
On the Day of the Full Moon

On April 25, 1986, the *Albany Times-Union* told of Eric Paquette's death. It said that "Friend fires in 'horseplay'." Paquette, of Middleburgh, NY, died when he and his 16-year-old friend were alone in an upstairs room of his friend's home. It was reported that they had decided to play with the weapons in the gun cabinet. The longitude and latitude for Middleburgh, NY are 074W20 and 42N36.

Boy, 13, killed by gunshot

Friend fires in 'horseplay'

By Joe Mahoney
Staff writer

MIDDLEBURGH – A lock on a cabinet failed to stop two teenag Thursday from gaining access supply of rifles – one of whic accidentally fired by one of the killing his companion, authorities sai

The tragic death of 13-year-old D. Paquette of Bush Road in Schoharie County village came on heels of another tragic shooting t took the life of a 4-year-old Ul County boy Wedenesday.

Illustration 22 - Albany Times Union (Paquette)

Paquette's friend was holding a .22-caliber rifle a few feet away from him when the weapon discharged, fatally striking him in the left-side of the chest. The 16 year-old then took Paquette downstairs and out onto the front porch of his home, where two or three other teenage boys had reportedly been loitering. Paquette's death occurred on April 24, the day of the full moon and the total eclipse.

Ironically, less than six months after Paquette's death, the *Schenectady Gazette* featured the following stories about a Satanic cult in the Middleburgh area:

"Says Middleburgh Pastor--"Don't Overreact to Rumors of a Satanic Cult," September 24, 1986.
"Rumors Often Enlarged By Anxiety, Uncertainty," September 29, 1986.
"DA Airs Cult Rumors, Says Charges Probable," September 30, 1986.
"Middleburgh High Fills For "Satanic Cult" Talk," October 1, 1986.

"Book Tells of 1914 Outbreak--Police, Teachers Say Cult Rumors Decreasing," October 2, 1986.
"Small Animal Sacrifice Note Gets Man Jailed," October 18, 1986.
"1986 in Review--Tales of Satanism Contend for Top Scoharie County Story," December 29, 1986.

It had been rumored that there was a devil cult operating in Scoharie County long before the shooting of Eric Paquette. The rumors continued all that summer. In September, when the above articles were published in the newspapers, people's fears had reached a state of panic. It was thought that as many as 30 to 50 teenagers and young adults might be members of a devil cult.

As a parapsychologist, the writer was consulted on the cult situation. Town officials asked whether the teenagers and young adults had psychic powers or if their cult was drug-related. Officials had also heard rumors of sacrificial ceremonies to be scheduled around Halloween. In an effort to end further devil cult rumors, the town officials made it known they were looking into possible criminal charges that might or might not have had cult overtones. It was further stated, "The charges will be for something other than just writing graffiti."

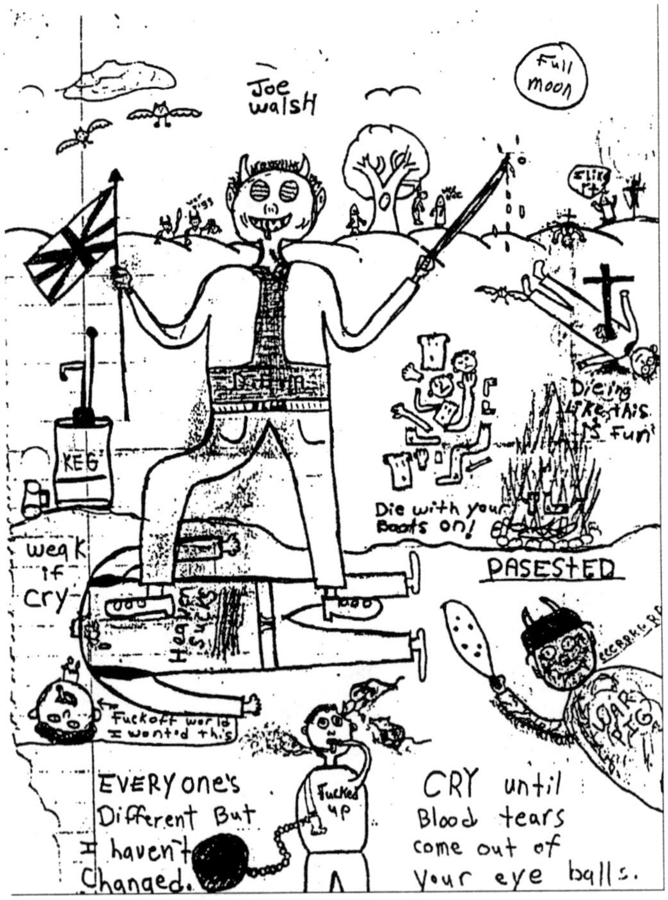

Illustration 23 - Drawing by another boy possibly depicting Paquette's death from bullet in right chest and during the full moon (Paquette)

Suddenly, the cult rumors, incidences and graffiti in Middleburgh and Scoharie County ended. However, one might remember the young boy who died on the day of the full moon and the total eclipse. Was his death cult related, as it had been rumored? Hand-drawn sketches by another boy in jail for drugs shortly after Paquette's death depict cult related beliefs and possibly even his death.

Three interpretations from Paquette's astrological chart which give us an indication of the influences he was under at the time of his death. These interpretations are in part from *Predictive Astrology* by Frances Sakoian and Louis Acker or *Modern Transits* by Lois Rodden.

Uranus ssq Jupiter (A) -- One may come into contact with disturbing religious and cultural concepts or become associated with cults and organizations of a dubious nature. This is not a period to indulge in strange religious and psychic practices. An unfavorable time for psychic and occult experimentation.
Uranus sxt Uranus (A) -- Stimulating new ideas. Some people develop an interest (possibly the occult?). Unusual friends and organizations.
Pluto ses Mercury (A) -- Unpleasant psychic disturbances. May become mentally preoccupied with death, life after death, reincarnation or occult subjects.

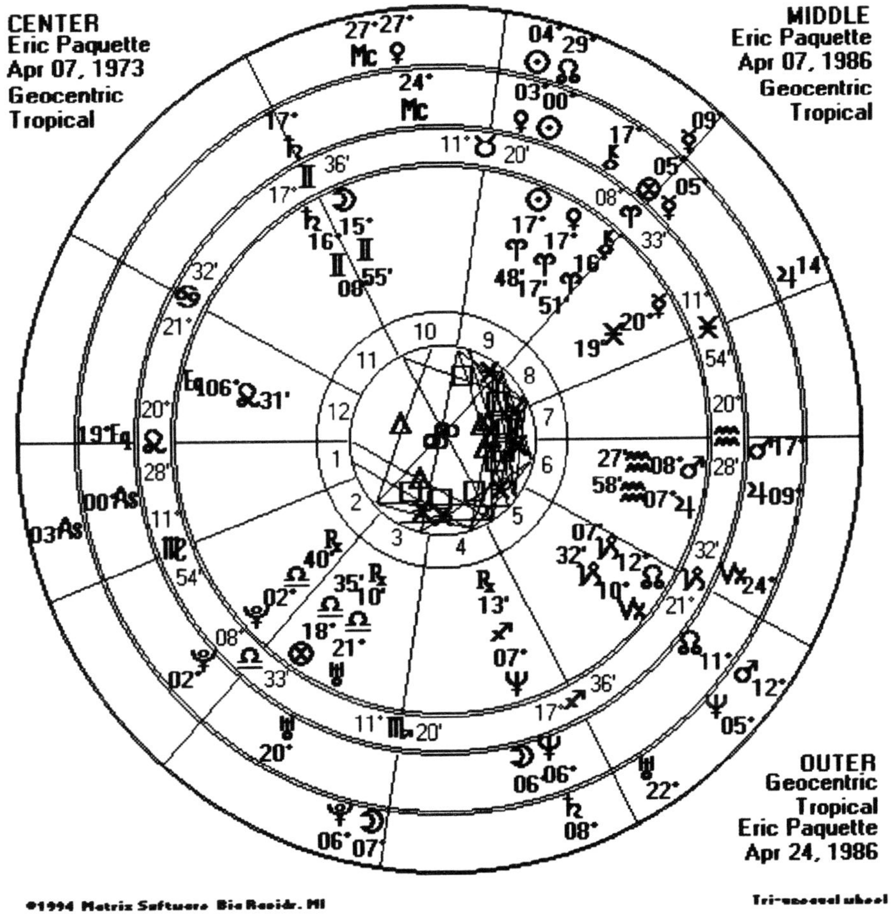

Chart 40 - Eric Paquette's Three-ring Chart with Natal, Progressed Chart for 1986 and Transits for April 24, 1986

Paquette's Natal Chart

In Paquette's natal chart there are no critical degrees, intercepted signs, yods or 29/00 degree placements. His part of Fortuna at 18:35 Libra in the third house of friends and everyday environment opposes the Sun within one degree at 17:48 Aries in the ninth house of abstract thought. Mars at 08:27 Aquarius in the sixth house of health and service to others conjuncts Jupiter at 07:58 Aquarius. The Sun at 17:48 Libra closely conjuncts Venus at 17:17 Aries both in the ninth house. These two planets form sextiles to the Moon at 15:55 Gemini and Saturn at 16:08 Gemini, both in the tenth house of status and public recognition.

The Progressed Chart for 1986

There is a grand trine in Paquette's progressed chart comprised of progressed Uranus at 20:37 Libra R in the natal third house of friends and everyday environment, progressed Mars at 17:41 Aquarius in the natal sixth house of health and service to others, and Saturn at the critical degree of 17:23 Gemini in the natal tenth house of public recognition. There is also a wide yod comprised of progressed Pluto at 02:21 Libra R in the second house of finances and values, the progressed Moon at 06:11 Sagittarius and Neptune at 06:59 Sagittarius R, both in the natal fourth house of home and personal matters, and the progressed Sun at 00:32 Taurus and Venus at 03:22 Taurus, both in the natal ninth house of abstract mind.

Progressed Jupiter is at the critical degree of 09:41 Aquarius in the sixth house of health and service to others. It forms no close aspects. The progressed Moon at 06:11 Sagittarius and Neptune at 06:59 Sagittarius R, both in the natal fourth house of home, trine both progressed Mercury at 05:07 Aries and Fortuna or Misfortuna at 05:58 Aries in the natal eighth house of shared resources.

The Total Eclipse and the Full Moon

The Total Eclipse of Moon and the full moon occurred on April 24, 1986 at 04:37 Scorpio. The eclipse falls in Paquette's third house and can indicate changes in mental attitudes, friends and everyday environment. Sophia Mason in *Forecasting with the New, Full & Quarter Moons Through the Houses* interprets a third house eclipse as "Possible separation from close relatives or trouble with neighbors."

Solar Return Chart for 1986 (Middleburgh, NY)

There in a grand trine in Paquette's solar return chart comprised of Pluto at 06:29 Scorpio R in the solar second house of values, Jupiter at 10:43 Pisces in the sixth house of health, and the solar Midheaven at 06:24 Cancer. There is also a grand cross comprised of the solar Ascendant and first house of self at 05:26 Libra, Mars at 05:17 Capricorn and Neptune at 05:49 Capricorn R, both in the third house of friends and everyday environment, the Moon at 02:42 Aries in the sixth house of health, and the Midheaven at 06:24 Cancer.

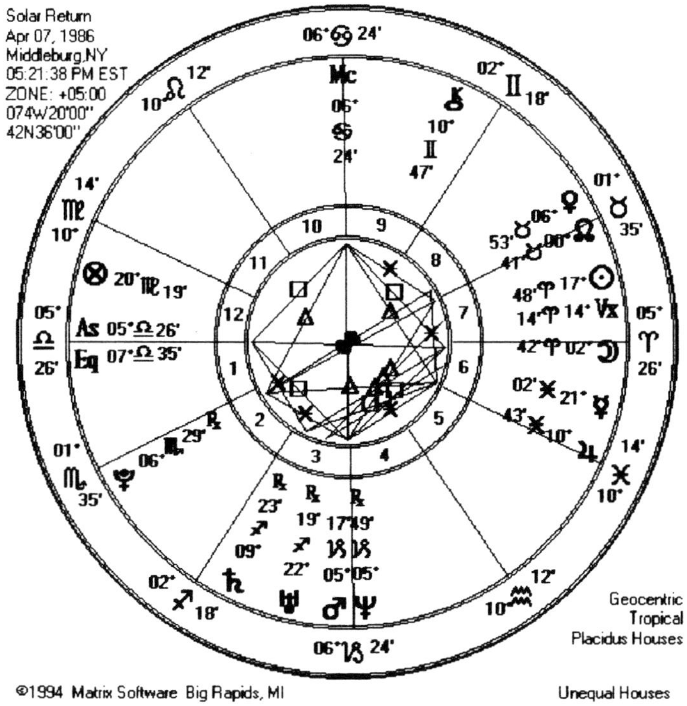

Chart 41 - Eric Paquette's Solar Return Chart for 1986

In Conclusion

No one was charged in Eric's death and it was ruled as accidental.

Birth Data for Eric Paquette

Birth date:	April 7, 1973
Birth time:	01:30 P.M./E.S.T.
Birth place:	Cobleskill, New York
Longitude:	074W29 Latitude: 42N41
Source:	From mother

Placements and degrees of Paquette's charts:
Natal Placements

Moon	15:55 Gem	10th	Neptune	07:13 Sag R	04th
Sun	17:48 Ari	09th	Pluto	02:40 Lib R	02nd
Mercury	20:19 Pis	08th	North Node	12:07 Cap R	05th
Venus	17:17 Ari	09th	Vertex	10:32 Cap	05th
Mars	08:27 Aqu	06th	East Point	06:31 Leo	12th
Jupiter	07:58 Aqu	06th	Fortuna	18:35 Lib	03rd
Saturn	16:08 Gem	10th	Midheaven	11:20 Tau	
Uranus	21:10 Lib R	03rd	Ascendant	20:28 Leo	

Eleventh house position is 17:36 Gem
Twelfth house position is 21:32 Tau
Second house position is 11:54 Vir
Third house position is 08:33 Lib

Critical Degrees: None
To be noted: None
Intercepted Signs: None

Progressed Placements for 1986

Moon	06:11 Sag	04th	Neptune	06:59 Sag R	04th
Sun	00:32 Tau	09th	Pluto	02:21 Lib R	02nd
Mercury	05:07 Ari	08th	North Node	11:26 Cap R	05th
Venus	03:22 Tau	09th	Vertex	24:57 Cap	06th
Mars	17:41 Aqu	06th	East Point	19:16 Leo	12th
Jupiter	09:41 Aqu	06th	Fortuna	05:58 Ari	08th
Saturn	17:23 Gem	10th	Midheaven	24:04 Tau	10th
Uranus	20:37 Lib R	03rd	Ascendant	00:18 Vir	01st

Critical Degrees: Jupiter, Saturn
To be noted: Sun at 00 Taurus in the ninth house of the abstract mind, and the Ascendant at 00 Virgo

Transiting Placements for April 24, 1986

Moon	07:36 Sco	03rd	Uranus	22:03 Sag R	05th
Sun	04:17 Tau	09th	Neptune	05:44 Cap R	05th
Mercury	09:22 Ari	09th	Pluto	06:01 Sco	03rd
Venus	27:32 Tau	10th	North Node	29:48 Ari R	09th
Mars	12:42 Cap	05th	Vertex	29:50 Cap	06th
Jupiter	14:10 Pis	08th	East Point	23:10 Leo	01st
Saturn	08:40 Sag R	04th	Fortuna	06:34 Pis	07th

Critical Degrees: None
To be noted: North Node at 29 degrees of Taurus in the ninth house of abstract thought and Vertex at 29 degrees of Capricorn in the sixth house of physical health

Solar Return Placements for 1986 (Middleburgh, NY)

Moon	02:42 Ari	06th	Neptune	05:49 Cap R	03rd
Sun	17:48 Ari	07th	Pluto	06:29 Sco R	02nd
Mercury	21:02 Pis	06th	North Node	00:41 Tau R	07th
Venus	06:53 Tau	08th	Vertex	14:14 Ari	07th
Mars	05:17 Cap	03rd	East Point	07:35 Lib	01st
Jupiter	10:43 Pis	06th	Fortuna	20:19 Vir	12th
Saturn	09:23 Sag R	03rd	Midheaven	06:24 Can	

Uranus	22:19 Sag R	03rd	Ascendant	05:26 Lib

Eleventh house position is 10:12 Leo
Twelfth house position is 10:14 Vir

Second house position is 01:35 Sco
Third house position is 02:18 Sag

Critical Degrees: None
To be noted: North Node at 00 degrees in the seventh house of one-to-one relationships
Intercepted Signs: None

Delores Davidson
Was it Suicide?

Delores Davidson left her home in the Flowing Wells area of Tucson, AZ during the early afternoon of June 15, 1983 to stop at the bank to cash a check for $30 before going to work. She was found two days later 35 miles east of San Diego, which is more than 400 miles away. Davidson was slumped over the wheel of her parked car, shot once in the forehead. The pistol was in her right hand, which showed gunpowder burns. There was no sign of a struggle or rape. Even though there was no suicide note, homicide ruled her death a suicide. Her husband disagreed with the ruling.

Picture 9 - Delores Davidson

Davidson's husband, Bill, a former policeman and reporter, believed that his wife was abducted and that her so-called suicide was staged. He was suspicious because the passenger side door of the car was unlocked. Bill Davidson said that was unusual because his wife always kept all the doors locked.

At the time of her death, Davidson's daughter was living with her and Bill, her second husband. Problems existed in the mother and daughter's relationship because of the daughter's recent series of criminal charges for bad checks, child neglect and drugs. Davidson also did not approve of her daughter's boyfriend because he had a criminal record.

Bill Davidson mounted a one-man inquiry into his wife's death, but could come up with no conclusive evidence to change the ruling of suicide. He felt that there were significant factors favoring homicide. First, Davidson had filled her car with gas for about $13 and the remaining $17 was missing. In addition, the right car door was unlocked, and lastly, Davidson had not prepared a will.

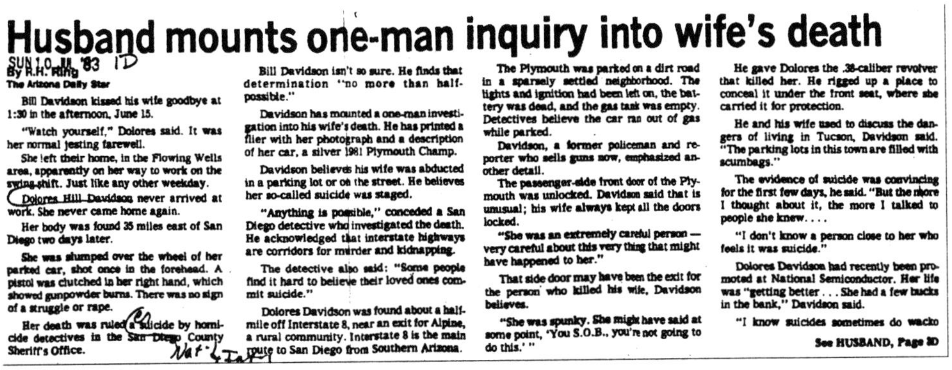

Husband mounts one-man inquiry into wife's death

SUN 10 JUL '83 1D

By R.H. Ring
The Arizona Daily Star

Bill Davidson kissed his wife goodbye at 1:30 in the afternoon, June 15.

"Watch yourself," Dolores said. It was her normal jesting farewell.

She left their home, in the Flowing Wells area, apparently on her way to work on the swing shift. Just like any other weekday.

Dolores Hill Davidson never arrived at work. She never came home again.

Her body was found 35 miles east of San Diego two days later.

She was slumped over the wheel of her parked car, shot once in the forehead. A pistol was clutched in her right hand, which showed gunpowder burns. There was no sign of a struggle or rape.

Her death was ruled a suicide by homicide detectives in the San Diego County Sheriff's Office.

Bill Davidson isn't so sure. He finds that determination "no more than half-possible."

Davidson has mounted a one-man investigation into his wife's death. He has printed a flier with her photograph and a description of her car, a silver 1981 Plymouth Champ.

Davidson believes his wife was abducted in a parking lot or on the street. He believes her so-called suicide was staged.

"Anything is possible," conceded a San Diego detective who investigated the death. He acknowledged that interstate highways are corridors for murder and kidnapping.

The detective also said: "Some people find it hard to believe their loved ones commit suicide."

Dolores Davidson was found about a half-mile off Interstate 8, near an exit for Alpine, a rural community. Interstate 8 is the main route to San Diego from Southern Arizona.

The Plymouth was parked on a dirt road in a sparsely settled neighborhood. The lights and ignition had been left on, the battery was dead, and the gas tank was empty. Detectives believe the car ran out of gas while parked.

Davidson, a former policeman and reporter who sells guns now, emphasized another detail.

The passenger-side front door of the Plymouth was unlocked. Davidson said that is unusual; his wife always kept all the doors locked.

"She was an extremely careful person — very careful about this very thing that might have happened to her."

That side door may have been the exit for the person who killed his wife, Davidson believes.

"She was spunky. She might have said at some point, 'You S.O.B., you're not going to do this.' "

He gave Dolores the .38-caliber revolver that killed her. He rigged up a place to conceal it under the front seat, where she carried it for protection.

He and his wife used to discuss the dangers of living in Tucson, Davidson said. "The parking lots in this town are filled with scumbags."

The evidence of suicide was convincing for the first few days, he said. "But the more I thought about it, the more I talked to people she knew. . . .

"I don't know a person close to her who feels it was suicide."

Dolores Davidson had recently been promoted at National Semiconductor. Her life was "getting better . . . She had a few bucks in the bank," Davidson said.

"I know suicides sometimes do wacko

See HUSBAND, Page 3D

Illustration 24 - The Arizona Daily Star (Davidson)

Davidson's Natal Chart

In Davidson's natal chart, Libra is intercepted in the fifth house of children, pleasures and creativity, and Aries is intercepted in the eleventh house of friends, acquaintances, hopes and wishes. Davidson had three children from her first marriage. Neptune at the critical degree of 00:31 Libra conjuncts her fifth house cusp of 29:34 Virgo.

The Sun at 26:24 Leo in the third house of everyday environment opposes the Midheaven at 27:27 Aquarius and is the aspect with the tightest orb.

The Progressed Chart for 1982

Davidson's progressed Sun at 04:18 Libra in the natal fifth house of children, pleasures and creativity opposes the progressed Midheaven at 05:20 Aries, and sextiles progressed Venus at the critical degree of 04:22 Virgo in the natal fourth house of domestic and personal matters. Progressed Fortuna at 05:29 Cancer in the natal first house of self also sextiles progressed Venus at 04:22 Virgo in the fourth house and squares the progressed Sun at 04:18 Libra in the natal fifth house.

Davidson's Transits Interpreted for June 15, 1983

Transiting Venus is at the critical degree of 09:00 Leo in the natal third house of everyday environment, and transiting Pluto is at the critical degree of 26:51 Libra R in the natal fifth house of children and love relationships. Transiting Jupiter at 03:48 Sagittarius is retrograding from the critical degree of 04 Sagittarius in the natal sixth house of health and service.

There are three aspects of the outer planets within a close. Transiting Saturn at 27:56 Libra R in the natal fifth house retrogrades back to an inconjunct with natal Mars at 27:54 Taurus in the natal twelfth house of secret enemies and self-undoing. The exact date of this conjunction is June 16, 1983. Transiting Neptune at 27:59 Sagittarius R in the natal seventh house of one-to-one relationships retrogrades back to inconjunct natal Mars at 27:54 Taurus in the natal twelfth house. The exact inconjunction falls on June 18, 1983. Transiting Neptune at 27:59 Sagittarius R in the natal seventh house retrogrades back to trine the natal Moon at 27:35 Aries in the natal eleventh house of friends, hopes and wishes. This exact trine occurred on June 29, 1983.

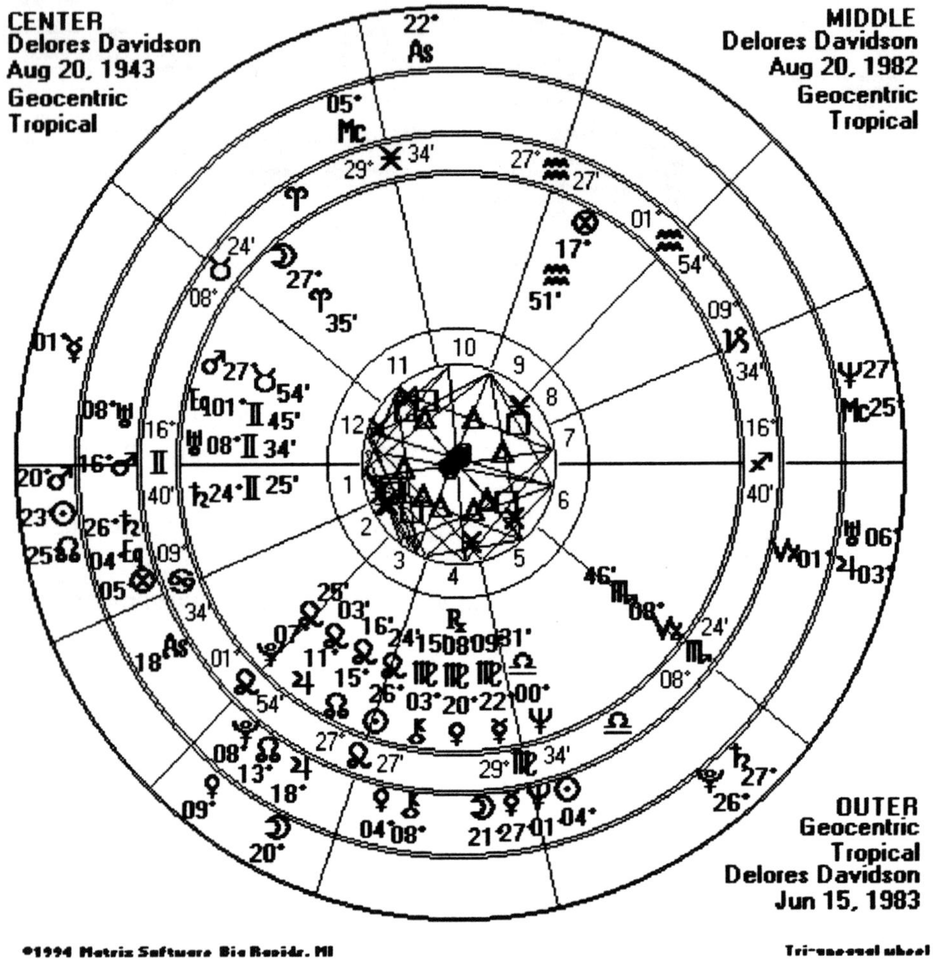

Chart 42 - Delores Davidson's Three-ring Chart with Natal, Progressed Chart for 1982 and Transits for June 15, 1983

Lois Rodden, writer of *Modern Transits*, interprets these transiting aspects as follows:

Saturn inc Mars (A) -- (Interpretation for square) A restriction, loss or trauma calls for hard work and disciplined energy.

Neptune inc Mars (A) -- (Interpretation for square) This is a potentially tricky or dangerous time that takes integrity and self-control to keep your ideals.

Neptune tri Moon (A) -- This is an idyllic domestic period, with more ease, camaraderie and fun.

A Partial Solar Eclipse on January 25, 1982 occurred at 04:53 Aquarius. The eclipse falls in Davidson's natal ninth house and can indicate changes in abstract mind and legal matters.

The Solar Return Chart for 1982 (Tucson, AZ)

Davidson's solar return chart for 1982 is a bowl-type chart with Capricorn intercepted in the solar third house of everyday environment, and Cancer intercepted in the solar ninth house of abstract mind and legal matters. The solar Moon at the critical degree of 04:53 Virgo in the solar eleventh house of friends and acquaintances sextiles solar Jupiter at 04:19 Scorpio in the solar first house of self.

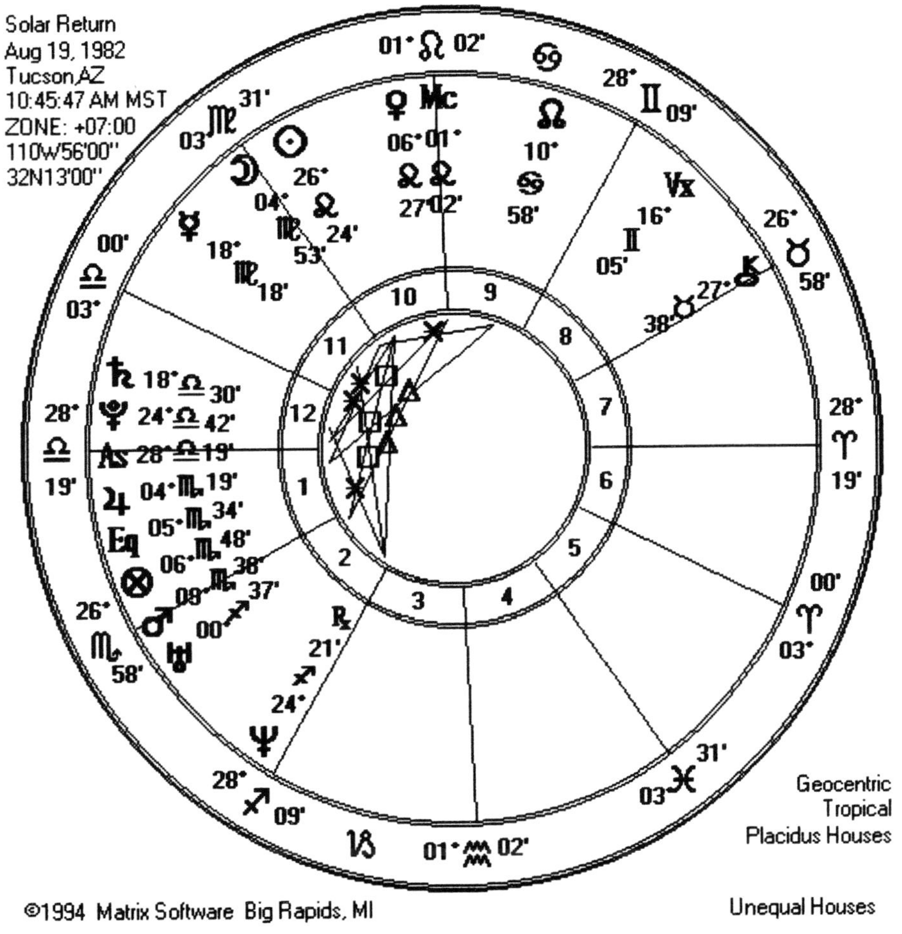

Chart 43 - Delores Davidson's Solar Return Chart for 1982

Solar Mars at the critical degree of 09:38 Scorpio in the solar first house of self trines the solar North Node at 10:58 Cancer R in the solar ninth house of abstract mind and legal matters. Solar Uranus at 00:37 Sagittarius in the solar second house of finances trines the Midheaven at 01:05 Leo. Neptune at 24:21 Sagittarius R in the second house of finances sextiles solar Pluto at 24:42 Libra in the solar twelfth house of secret enemies. Solar Venus at 06:27 Leo in the solar tenth house of public recognition squares solar Fortuna/Misfortuna at 06:48 Scorpio in the solar first house of self.

The Yod or Fingers of Fate

There is one yod that should be noted in Davidson's combined charts. It is comprised of transiting Pluto at 26:51 Libra and Saturn at 27:56 Libra R, both in the natal fifth house of children and romantic relationships, and the intercepted sign of Libra, transiting Neptune at 27:59 Sagittarius R in the natal seventh house of one-to-one relationships and natal Mars at 27:54 Taurus in the natal twelfth house of secret enemies.

The Yod

Transiting Pluto	Fifth House	26:51 Libra
Transiting Saturn	Fifth House	27:56 Libra R
Transiting Neptune	Seventh House	27:59 Sagittarius R
Natal Mars	Twelfth House	27:54 Taurus

In Conclusion

Delores Davidson's death was ruled as a suicide.

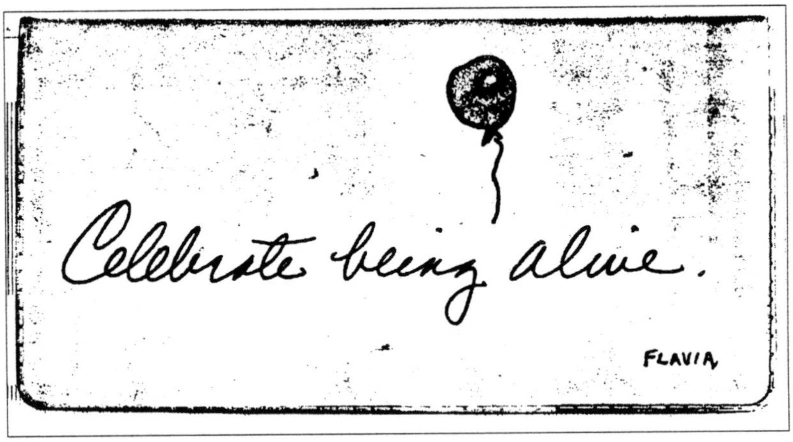

Illustration 25 - Davidson's checkbook cover (Davidson)

Birth Data for Delores Davidson:

Birth date:	August 20, 1943		
Birth time:	01:25 AM/MWT		
Birth place:	Morenci, AZ		
Longitude:	109W22	Latitude:	33N05
Source:	From family		

Placements and degrees of Davidson's charts:
Natal Placements

Moon	27:35 Ari	11th	Neptune	00:31 Lib	05th
Sun	26:24 Leo	03rd	Pluto	07:25 Leo	03rd
Mercury	22:09 Vir	04th	North Node	15:16 Leo R	03rd
Venus	20:08 Vir R	04th	Vertex	08:46 Sco	06th
Mars	27:54 Tau	12th	East Point	01:45 Gem	12th
Jupiter	11:03 Leo	03rd	Fortuna	17:51 Aqu	09th
Saturn	24:25 Gem	01st	Midheaven	27:27 Aqu	
Uranus	08:34 Gem	12th	Ascendant	16:40 Gem	

Eleventh house position is 29:34 Pis
Twelfth house position is 08:24 Pis
Second house position is 09:34 Can
Third house position is 01:54 Leo

Critical Degrees: Neptune
To be noted: Vertex approaching the critical degree of 09 Scorpio
Intercepted Signs: Libra intercepted in the fifth house of children, pleasures and creativity, and Aries intercepted in the eleventh house of friends, acquaintances, hopes and wishes

Progressed Placements for 1982

Moon	21:03 Vir	04th	Neptune	01:54 Lib	05th
Sun	04:18 Lib	05th	Pluto	08:21 Leo	03rd
Mercury	27:10 Vir R	04th	North Node	13:12 Leo R	03rd
Venus	04:22 Vir	04th	Vertex	01:53 Sag	06th
Mars	16:34 Gem	12th	East Point	04:30 Can	01st
Jupiter	18:59 Leo	03rd	Fortuna	05:29 Can	01st
Saturn	26:31 Gem	01st	Midheaven	05:20 Ari	11th
Uranus	08:46 Gem R	12th	Ascendant	18:43 Can	02nd

Critical Degrees: Venus
To be noted: None

Transiting Placements for June 15, 1983

Moon	20:12 Leo	03rd	Uranus	06:25 Sag R	06th
Sun	23:39 Gem	01st	Neptune	27:59 Sag R	07th
Mercury	01:16 Gem	12th	Pluto	26:51 Lib R	05th
Venus	09:00 Leo	03rd	North Node	25:06 Gem R	01st
Mars	20:29 Gem	01st	Vertex	26:40 Vir	04th
Jupiter	03:48 Sag R	06th	East Point	24:27 Pis	10th
Saturn	27:56 Lib R	05th	Fortuna	18:49 Tau	12th

Critical Degrees: Venus and Pluto
To be noted: Jupiter approaching the critical degree of 04 Sagittarius

Solar Return Placements for 1982 (Tucson, AZ)

Moon	04:53 Vir	11th	Neptune	24:21 Sag R	02nd
Sun	26:24 Leo	10th	Pluto	24:42 Lib	12th
Mercury	18:18 Vir	11th	North Node	10:58 Can R	09th
Venus	06:27 Leo	10th	Vertex	16:05 Gem	08th
Mars	09:38 Sco	01st	East Point	05:34 Sco	01st
Jupiter	04:19 Sco	01st	Fortuna	06:48 Sco	01st
Saturn	18:30 Lib	12th	Midheaven	01:02 Leo	
Uranus	00:37 Sag	02nd	Ascendant	28:19 Lib	

Eleventh house position is 03:31 Vir
Twelfth house position is 03:00 Lib
Second house position is 26:58 Sco
Third house position is 28:09 Sag

Critical Degrees: Moon, Mars
To be noted: Uranus at 00 Sagittarius
Intercepted Signs: Capricorn intercepted in the third house of everyday environment, and Cancer intercepted in the ninth house of abstract mind and legal matters

Joseph Cohen, Jr.
A Man with a Future

Joseph Cohen, Jr. was found hanging from a tree behind Our Lady of Angels Cemetery in Colonie, NY at about 7:00 A.M. on April 30, 1992. Police immediately ruled his death a suicide, even though there was no suicide note. However, it will be reminded that April 29, 1992 was commonly known as the Rodney King Riot Night, and Joseph was a young black man whose best friend and girlfriend were white. Cohen's parents firmly believe that their son's death was not a suicide. The longitude and latitude for Colonie, NY are 73W48 and 42N44.

Illustration 26 - Ad that appears regularly in the Albany Times Union (Cohen)

Cohen worked as a graphic artist for Kinko, Inc., a national copy and desk-top publishing company. At the time of his death, Cohen's future looked promising. He was about to open his own desk-top publishing business and had been to New York City just the week before to purchase supplies for the new business. It was suggested that Cohen may have had a disagreement with his employer because he was hesitant to sign a paper agreeing not to take any of the firm's customers. However, he got along well with his co-workers and had dinner with one that evening, and then returned to work at about 8:45 P.M.

The circumstances surrounding Cohen's death remain suspicious. The police originally theorized that because his 1988 Honda sedan was found in the nearby Colonie Shopping Center with its passenger-side crashed into a pole, he had tried to kill himself by slamming into the pole. The police later decided that this was improbable because an accident reconstruction determined that the vehicle's speed on impact was ten to fifteen miles per hour. Some still question whether Cohen was trying to flee from an assailant because the car was found with the driver's door open.

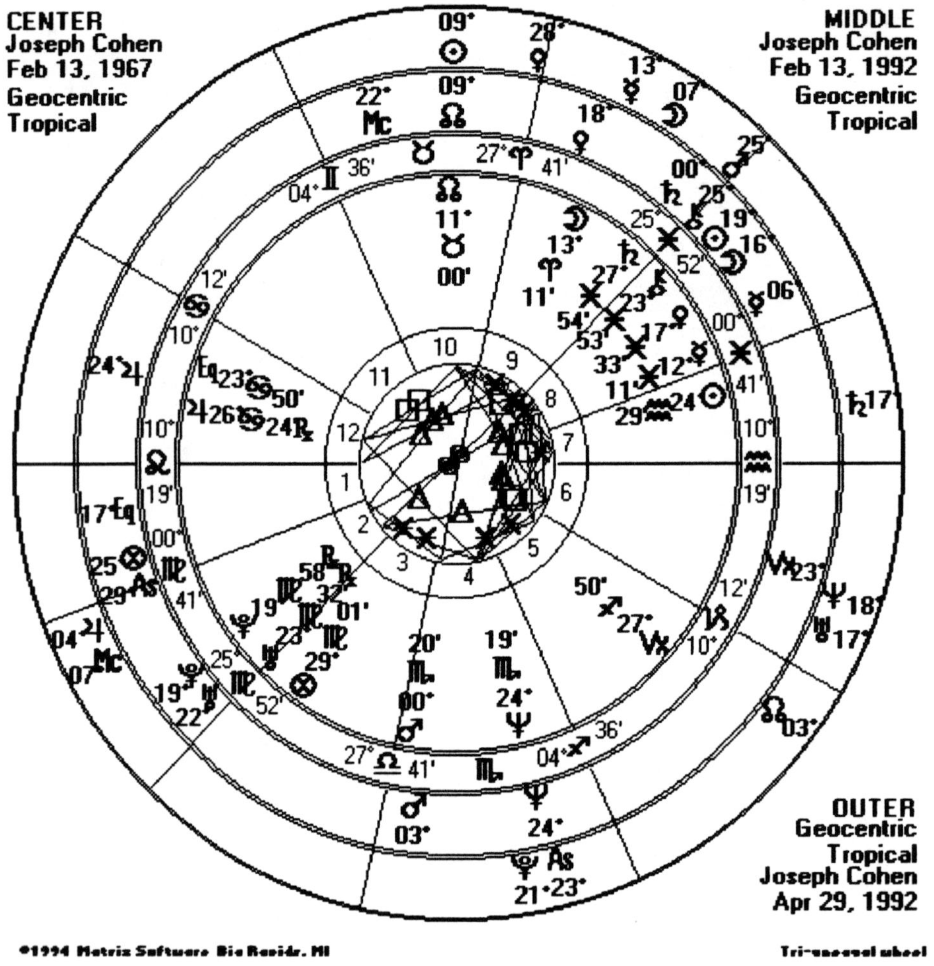

Chart 46 - Joseph Cohen's Three-ring Chart with Natal, Progressed Chart for 1992 and Transits for April 29, 1992

The police also theorize that when his first attempt at suicide failed, Cohen walked a distance to the cemetery. He went over a fence and through some brush, then found a tree and tried to hang himself again with his tie and belt. When that failed, he walked to a school parking lot where he found a truck with a thick hemp rope on the back. Cohen did not have a knife to cut off a piece of the rope, and it is not known what was used to cut it. He walked back to the tree with the piece of rope and successfully hung himself.

Several of the facts are questionable because of their inconsistency. (1) The police state the ground under the tree had not been disturbed. (2) Except for the box-cutter that Cohen had neatly tucked away in his wallet which had not been used to cut the rope, the police found no knife or cutting device. (3) Despite the walking and theorized suicide attempts, Cohen was found with his clothes clean and untouched, his glasses still secure on his head. (4) The car was not dusted for finger prints. (5) The police found three drugs in Cohen's car. Two were Immodium and Methotrexate, which he took for colitis. The other was an anti-depressant, Largactil, which the Cohens said their son did not take. The police were never able to determine how Cohen came into possession of this medication of English manufacture, and there were no traces of it in his body.

Cohen's parents have had two investigators attempt to answer their unanswered questions. They ask: How did their son cut the rope used in the hanging? Where did he get the Largactil? Was his rectal bleeding caused by a sexual attack?

The autopsy reports the death as strangulation by ligature. There was an absence of bruises, abrasions or other markings on the body, and acute rectal hemorrhaging. In addition, Cohen's death has been featured on NBC's "*The Crusaders*," a show which investigates current issues. The broadcast findings supported the belief that his death was not a suicide. In addition, the Coalition of Concerned Citizens to Solve the Lynching of Joseph Cohen, Jr., raised several questions surrounding his death.

Cohen's Natal Chart

In Cohen's natal chart there is a grand trine comprised of Neptune at 24:19 Scorpio in the fourth house of home and early environment, Saturn at 27:54 Pisces in the ninth house of abstract mind and philosophies and Jupiter at the critical degree of 26:24 Cancer R in the twelfth house of self-undoing and secret enemies. There is also a yod comprised of Uranus at 23:32 Virgo R in the second house of finances and values, the Sun at 24:29 Aquarius in the seventh house of one-to-one relationships, and Jupiter at the critical degree of 26:24 Cancer in the twelfth house of secret enemies and self-undoing.

The Grand Trine

Neptune	Fourth House	24:19 Scorpio
Saturn	Ninth House	27:54 Pisces
Jupiter	Twelfth House	26:24 Cancer R

The Yod

Uranus	Second House	23:32 Virgo R
Sun	Seventh House	24:29 Aquarius
Jupiter	Twelfth House	26:24 Cancer R

The Moon at the critical degree of 13:11 Aries in the ninth house forms no close aspects. The Sun at 24:29 Aquarius in the seventh house of one-to-one relationships squares Neptune at 24:19 Scorpio in the fourth house of home and personal matters. Venus at the critical degree of 17:33 Pisces in the eighth house of shared resources opposes Pluto at 19:58 Virgo R in the second house of financial and personal values.

Taurus is intercepted in the tenth house of career and public recognition and Scorpio is intercepted in the fourth house of the emotional and physical home.

The Progressed Chart for 1992

The Ascendant in Cohen's progressed chart is at the decisive 29th degree of Leo. The progressed Sun at 19:36 Pisces in the natal eighth house of shared values and resources opposes Pluto at 19:20 Virgo R in the natal second house of personal values and resources. Progressed Venus at 18:22 Aries in the natal ninth house of abstract mind and publication inconjuncts Pluto at 19:20 Virgo R in the natal second house.

Progressed Saturn at the critical degree of 00:51 Aries in the natal ninth house inconjuncts the progressed Ascendant at 29:18 Leo in the natal first house. The North Node is at the critical degree of 09:40 Taurus R in the natal tenth house of career and public recognition.

Cohen's Transits Interpreted for April 29, 1992

There are three transiting planets at critical degrees. Transiting Sun is at the critical degree of 09:56 Taurus which is intercepted in the natal tenth house of career and public recognition. Transiting Mercury at the critical degree of 13:46 Aries is in the ninth house of abstract mind. Transiting Pluto retrogrades back to the critical degree of 21:54 Scorpio R in the natal fourth house of the emotional and physical home. The sign Scorpio is also intercepted in the natal fourth house.

There is one aspect within orb between a transiting planet and a natal. Transiting Uranus at 17:59 Capricorn R in the natal sixth house of work and health retrogrades back to form a sextile with natal Venus at 17:33 Pisces in the natal eighth house of shared resources. Lois Rodden, writer of *Modern Transits,* interprets this transiting aspect as follows:

> **Uranus sxt Venus (A)** -- This is an opportune time to stay with old secure patterns rather than to start a new social program or romance.

An Annular Eclipse on January 4, 1992 occurred at 13:51 Capricorn. The eclipse falls in Cohen's sixth house and can indicate changes in occupation and service to others.

Solar Return Chart for 1992

Cohen's solar return chart for 1992 has one yod. The yod is formed by solar Pluto at 22:55 Scorpio in the fourth house of home and early environment, solar Venus at

24:08 Capricorn in the sixth house of health and service, and the solar Moon at 23:20 Gemini in the tenth house of profession, employer and public recognition.

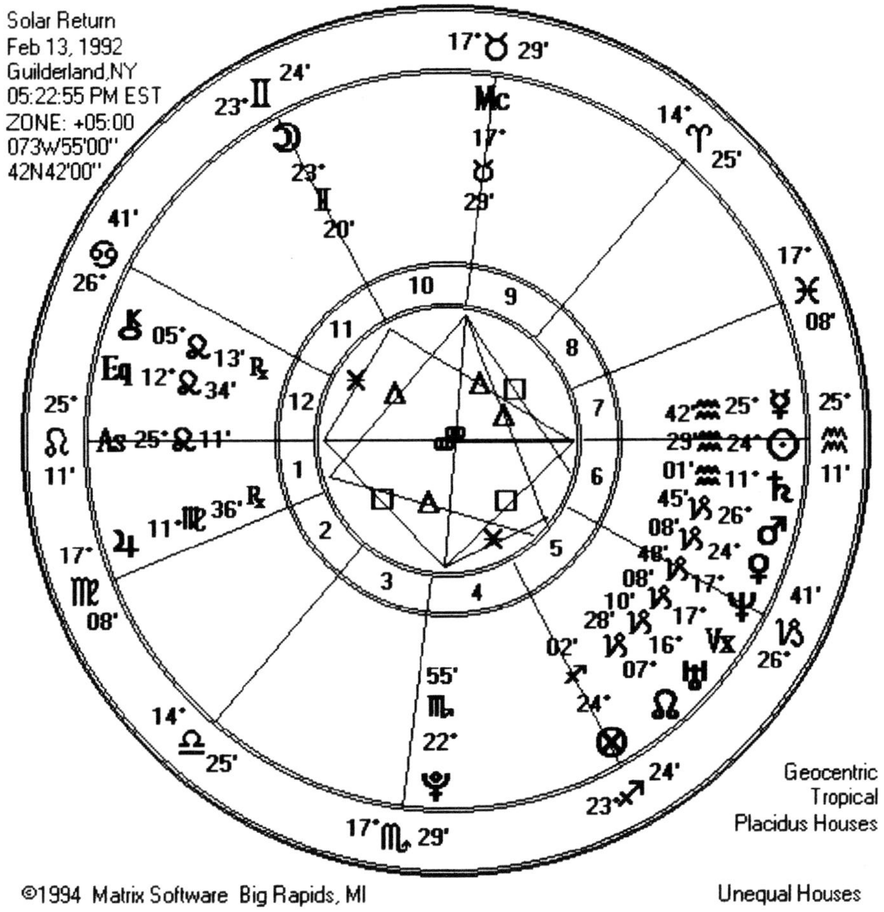

Chart 47 - Joseph Cohen's Solar Return Chart for 1992

The Yod

Pluto	Fourth House	22:55 Scorpio
Venus	Sixth House	24:08 Capricorn
Moon	Tenth House	23:21 Gemini

Solar Mars at the critical degree of 26:45 Capricorn in the sixth house of work and health forms only one aspect which is to the solar Ascendant at 25:11 Leo. The solar Sun at 24:29 Aquarius in the sixth house of occupation and service to others conjuncts solar Mercury at 25:42 Aquarius in the seventh house of one-to-one relationships. The solar Sun and Mercury oppose the solar Ascendant at 25:11 Leo.

Solar Jupiter at 11:36 Virgo R in the first house of self inconjuncts Solar Saturn at 11:01 Aquarius in the sixth house of health and occupation. Solar Uranus at 16:10 Capricorn conjuncts solar Neptune at 17:48 Capricorn in the solar fifth house of romantic relationships and they trine the solar Midheaven at 17:29 Taurus.

There is a fifth and sixth house stellium in Capricorn. The stellium is comprised of the North Node at 07:28 R, Uranus at 16:10, Neptune at 17:48, Venus at 24:08 and Mars at 26:45 all in Capricorn.

The Yod or Fingers of Fate

There is a yod is Cohen's combined charts comprised of the progressed Ascendant at the culminating 29:18 Leo in the first house of self, natal Mars at 00:20 Scorpio in the fourth house of home and personal matters, and progressed Saturn at the critical degree of 00:51 Aries in the ninth house of abstract mind.

The Yod

First House	Progressed Ascendant	29:18 Leo
Fourth House	Natal Mars	00:20 Scorpio
Ninth House	Progressed Saturn	00:51 Aries (Critical Degree)

The Yod includes the first, fourth and ninth houses

The Basic Keywords for this yod from Rex E. Bills *The Rulership Book* are divided into those which refer to Joseph and those which refer to his manner of death:

Joseph Cohen

One's actions	Ambitions, one's	Habits
Private life, one's private matters	One's character	Business, in general
	Self-opinion	Downfall, one's
Publications	Joseph	

Joseph Cohen's Death

Trees, in general	Aggressors, assault	Anger, arguments
Deaths by accident/violence	Crimes of violence	Conflict
Enemies, secret, profession	Agitators	Murder
Death, sudden	Competitors	Males, in general
Belts	Executions, executioners	Rectum
Mob fury, mob lynchings	Choking	Hangmen
Lynching, lynchings	Printing, printers	Hemp

In Conclusion

On April 11, 1994, the Albany County Legislature unanimously asked District Attorney Sol Greenberg to reopen an investigation into Cohen's death. Greenberg's statements said that there was no compelling reason to start a new investigation into Joseph Cohen's death because all the evidence says it was suicide. Greenberg said the case is not closed, just inactive. He further said, "If new evidence surfaces, I would consider reopening the case."

Birth Data for Joseph Cohen, Jr.:

Birth date:	February 13, 1967		
Birth time:	04:05 PM/EST		
Birth place:	Albany, NY		
Longitude:	073W47	Latitude:	42N39
Source:	From family		

Placements and degrees of Cohen's charts:
Natal Placements

Moon	13:11 Ari	09th	Neptune	24:19 Sco	04th
Sun	24:29 Aqu	07th	Pluto	19:58 Vir R	02nd
Mercury	12:11 Pis	08th	North Node	11:00 Tau R	10th
Venus	17:33 Pis	08th	Vertex	27:50 Sag	05th
Mars	00:20 Sco	04th	East Point	23:50 Can	12th
Jupiter	26:24 Can R	12th	Fortuna	29:01 Vir	03rd
Saturn	27:54 Pis	09th	Midheaven	27:41 Ari	
Uranus	23:32 Vir R	02nd	Ascendant	10:19 Leo	

Eleventh house position is 04:36 Gem
Twelfth house position is 10:12 Can
Second house position is 00:41 Vir
Third house position is 25:52 Vir

Critical Degrees: Moon, Venus, Jupiter
To be noted: Mars at 00 degrees Scorpio, Fortuna or Misfortuna at 29 degrees Virgo in the third house of friends and everyday environment
Intercepted Signs: Taurus in the tenth house of career and public recognition and Scorpio in the fourth house of personal and domestic affairs

Progressed Placements for 1992

Moon	16:13 Pis	08th	Neptune	24:18 Sco R	04th
Sun	19:36 Pis	08th	Pluto	19:20 Vir R	02nd
Mercury	06:56 Pis R	08th	North Node	09:40 Tau R	10th
Venus	18:22 Ari	09th	Vertex	23:21 Cap	06th
Mars	03:10 Sco R	04th	East Point	17:58 Leo	01st
Jupiter	24:36 Can R	12th	Fortuna	25:55 Leo	01st
Saturn	00:51 Ari	09th	Midheaven	22:49 Tau	10th
Uranus	22:30 Vir R	02nd	Ascendant	29:18 Leo	01st

Critical Degrees: Saturn, North Node
To be noted: Ascendant at 29 degrees Leo
Transiting Placements for April 29, 1992

Moon	07:14 Ari	09th	Uranus	17:59 Cap R	06th

Sun	09:56 Tau	10th	Neptune	18:56 Cap R	06th
Mercury	13:45 Ari	09th	Pluto	21:54 Sco R	04th
Venus	28:01 Ari	10th	North Node	03:27 Cap R	05th
Mars	25:30 Pis	08th	Vertex	06:43 Can	11th
Jupiter	04:38 Vir R	02nd	East Point	11:03 Sag	05th
Saturn	17:50 Aqu	07th	Fortuna	21:00 Lib	03rd

Critical Degrees: Sun, Mercury, Pluto
To be noted: None

Solar Return Placements for 1992 (Guilderland, NY)

Moon	23:20 Gem	10th	Neptune	17:48 Cap	05th
Sun	24:29 Aqu	06th	Pluto	22:55 Sco	04th
Mercury	25:42 Aqu	07th	North Node	07:28 Cap R	05th
Venus	24:08 Cap	05th	Vertex	17:08 Cap	05th
Mars	26:45 Cap	06th	East Point	12:34 Leo	12th
Jupiter	11:36 Vir R	01st	Fortuna	24:02 Sag	05th
Saturn	11:01 Aqu	06th	Midheaven	17:29 Tau	
Uranus	16:10 Cap	05th	Ascendant	25:11 Leo	

Eleventh house position is 23:24 Gem
Twelfth house position is 26:41 Can
Second house position is 17:08 Vir
Third house position is 14:25 Lib

Critical Degrees: Mars
To be noted: None
Intercepted Signs: None

Martin Brunolt
A Husband and Father Accused

This chapter about Martin Brunolt is being written because of a promise made to his mother. Brunolt is in prison for the 1994 murders of his wife and two children, but Marlene Chicoine believes in her son's innocence and wants his story told. As we look at what is known about the case, we can argue both his possible innocence or guilt.

The Brunolt family lived in Moreno Valley, CA. Brunolt, 45, married Graciela, 39, in 1984 and their marriage had produced two sons; Martin, Jr., 9, and Matthew, 6. Graciela was born in El Salvador and had become a naturalized citizen in 1992. She had a daughter by her first marriage who had lived with the Brunolts for the past seven years. Brunolt had also been married before and had three teenage daughters who were also living at their home. The longitude and latitude for Moreno Valley, CA is 117W15 and 33N56.

Illustration 27 - Brunolt Family Picture from Poster (Brunolt)

Graciela Brunolt did not speak English well, but she had worked her way up from fast-food restaurants and assembly lines to a supervisor for Kaiser Composites Tek. She was well liked by her co-workers and had attended church regularly. Brunolt worked as electrician's foreman for about seven years. However, he had recently been terminated from his job when he asked for a leave of absence because of personal problems.

Brunolt's personal problems over the past two years had stemmed from Graciela's affair with a co-worker. He filed for a divorce in May 1994, but dismissed the proceedings when Graciela ended her affair. At the time of the shootings, they were attending *Marriage Encounters* and working together to make a success of their reconciliation. Both had said that they were the happiest they had ever been.

The Brunolts plans for the future included opening a home-based business to end their hours of daily commuting. Additionally, Graciela felt that the children would be better off at home with her instead of their housekeeper/baby-sitter. They had begun working on expanding their home. Both the home expansion drawings and the loan had arrived the first week in August 1994. It was common knowledge amongst the family, housekeeper and possibly others that there would be sums of money in the house.

People who knew the Brunolts agree that they were not a family who lived in the realm of violence. However, on August 9, 1994, Brunolt was found unconscious and seriously injured from a bullet wound that had gone through his right jaw, his eye and the right frontal lobe of his brain. On the same night, he was arrested and accused of shooting to death his wife, Graciela, and their two sons, Martin, Jr., and Matthew. All of them, including Brunolt himself, had been shot once in the head.

Over the next month, Brunolt was kept isolated from his family until the judge ruled that they could visit. During their visit, his medical condition upset his family because he did not know them. He was not even aware of the fact that he was in prison. It was during that month that police claim Brunolt confessed three times to the murder of his family. However, copies of medical records show that at the same time he was on life support, taking morphine and his jaws had been wired shut.

Brunolt's medical care over the next ten months was questionable. Martin memory slowly began to return, but he had lost his vision and hearing on the right side. His first memory was of someone hitting him hard on the left side of the head as he entered the bedroom. He also remembers his wife trying to warn him of something, possibly that an intruder was in the room. Brunolt now believes there were a man and woman in the house at the time of the shooting.

Brunolt's Natal Chart

In Brunolt's natal chart Aquarius is intercepted in his second house of finances and money and Leo is intercepted in the eighth house of shared resources. Mercury is at the critical degree of 17:10 Gemini R in the sixth house of health and service to others, the North Node is at the critical degree of 23:52 Aries R in the fourth house of home and personal affairs, and the South Node at 23:52 Libra R is in the tenth house of public recognition and standing.

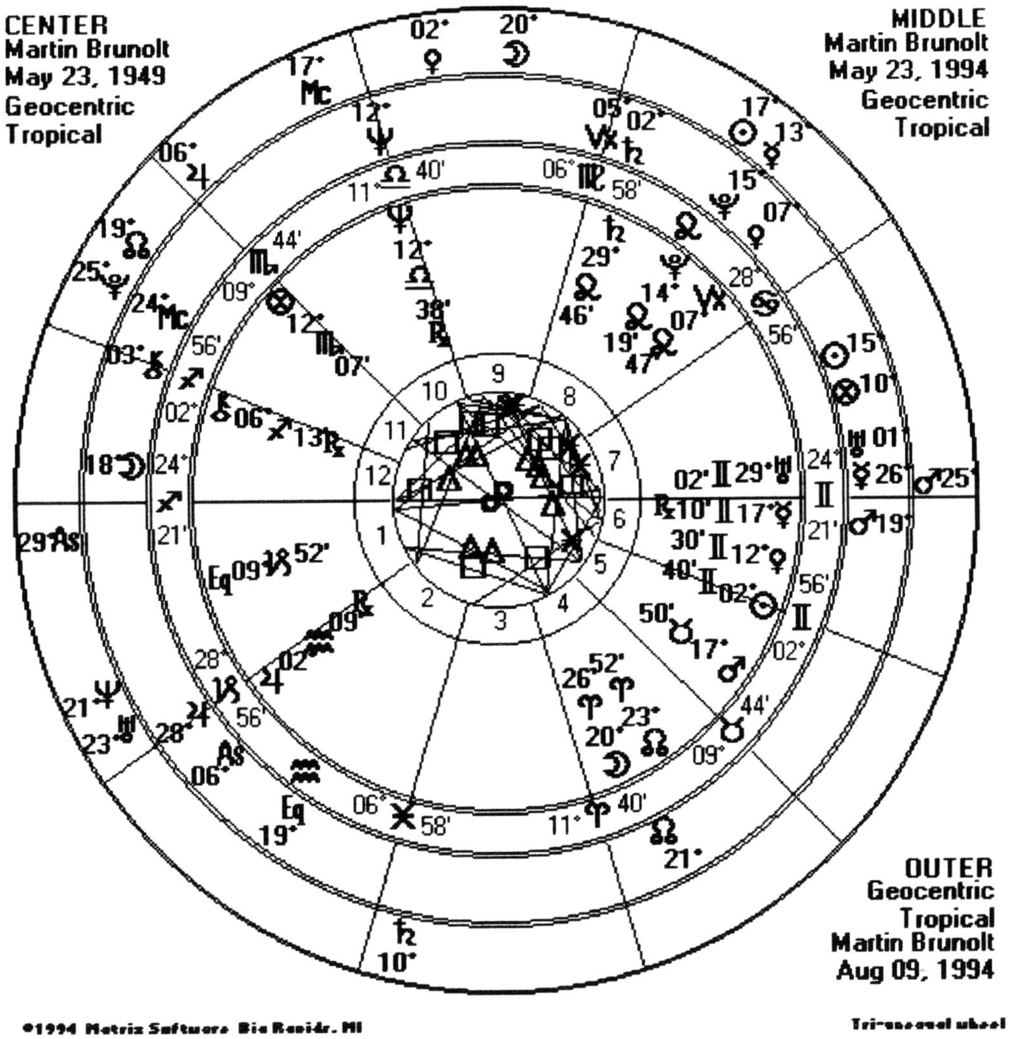

Chart 46 - Martin Brunolt's Three-ring Chart with Natal, Progressed Chart for 1994 and Transits for August 9, 1994

A grand trine is formed by the Ascendant 24:21 Sagittarius in the first house of self, North Node at 23:52 Aries R in the fourth house of personal and domestic affairs, and Saturn at 29:46 Leo in the eighth house of shared resources. Ordinarily, a five to six degree orb would not be used, but it is in this instance because of the 29 degrees and to bring attention to the houses involved.

As an added note of interest, Natal Pluto at 14:19 Leo in the eighth house of shared resources conjuncts the Part of Imprisonment and Captivity at 14:24 Leo.

The Progressed Chart for 1994

Progressed Jupiter has retrograded from the 29th degree of Capricorn in the natal second house back into the natal first house of self. The progressed Moon at 18:53 Sagittarius in the natal twelfth house of secret enemies opposes progressed Mars at 19:44 Gemini in the natal sixth house of service and others of service.

The Transiting Planets for August 9, 1994

There are no critical degrees in the transiting planets, or exact transits of the outer planets for August 1994. Transiting Saturn, Uranus and Neptune are in retrograde. In June, transiting Saturn in the natal third house of everyday environment in its retrograde motion moves away from a square to natal Venus in the natal sixth house of health, service to others and those who serve. Transiting Saturn also moves away from an inconjunct with natal Neptune in the natal tenth house of career and public recognition. In July, transiting Saturn forms an inconjunct with the natal Midheaven, and, thus, Martin's career problems began. Transiting Uranus in the natal first house of self squares the natal North Node in the natal fourth house of domestic and personal affairs on February 9, 1994, goes retrograde on May 2, 1994, and again squares the natal North Node on July 29, 1994. Uranus moves direct on October 2, 1994 and squares the natal North Node for the last time on December 1, 1994.

The Annular Solar Eclipse on May 10, 1994 at 19:49 Taurus took place in Martin's fifth house. A solar eclipse in the fifth house can indicate changes in connection with children, first child and romance.

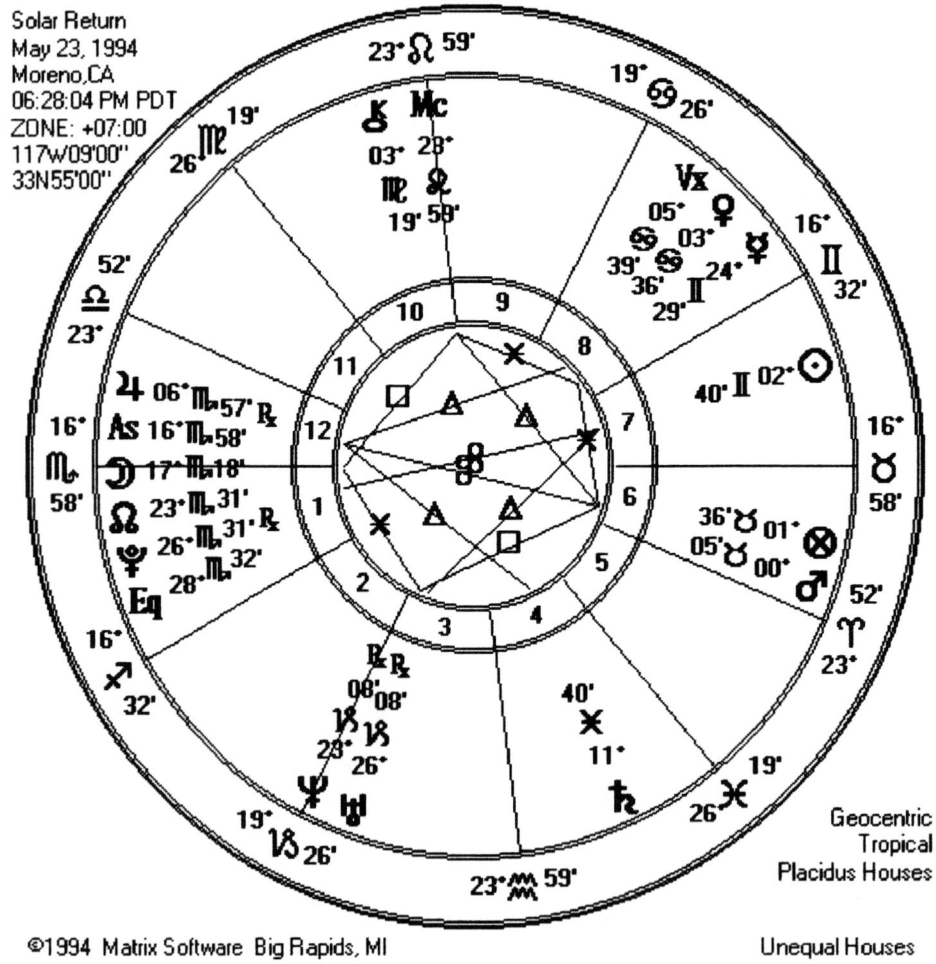

Chart 47 - Martin Brunolt's Solar Return Chart for 1994

The Solar Return Chart for 1994 (Moreno, CA)

Solar Uranus, the planet of sudden and unexpected events is at the critical degree of 26: 08 Capricorn R in the solar third house of everyday environment. Solar Mars is at 00:05 Taurus in the sixth house of health, service to others and others of service.

There are two yods in the solar return chart. One is comprised of solar Neptune at 23:08 Capricorn R and solar Uranus at 26:08 Capricorn R, both in the third house of everyday environment, solar Mercury at 24:29 Gemini in the eighth house of shared resources and values, and solar Midheaven at 23:59 Leo in the tenth house of career and public recognition.

The other yod is comprised on the solar North Node at 23:31 Scorpio R in the first house of self, solar Neptune at 23:08 Capricorn R and solar Uranus at 26:08 Capricorn R, both in the third house of everyday environment and solar Mercury at 24:29 Gemini in the eighth house of shared resources and values.

The Yod or Fingers of Fate

Two yods are found in the combined charts. The first yod is comprised of progressed Jupiter at 28:48 Capricorn R in the first house of self, natal Uranus at 29:02 Gemini in the seventh house of one-to-one relationships, and natal Saturn at 29:46 Leo in the eighth house of executions, death and matters of the dead.

The second yod is comprised of natal Moon at 20:26 Aries in the fourth house of domestic and personal affairs, progressed Mars at 19:44 Gemini in the sixth house of health and service of others, and transiting North Node at 19:23 Scorpio R in the eleventh house of friends and acquaintances.

The repetitive Basic Keywords from *The Rulership Book* by Rex E. Bills are used to interpret the yods in Brunolt's charts.

Yod No. 1

First House	Progressed Jupiter	28:48 Capricorn R
Seventh House	Natal Uranus	29:02 Gemini
Eighth House	Natal Saturn	29:46 Leo

The Yod includes the first, seventh and eighth houses

****

Yod No. 2

Fourth House	Natal Moon	20:26 Aries
Sixth House	Progressed Mars	19:44 Gemini
Eleventh House	Transiting Node	19:23 Scorpio R

The Yod includes the fourth, sixth and eleventh houses

The following is a list of the most repetitive words:

Brothers and sisters, in general
Friends of one's brothers and sisters
Arguments with public enemies
Brothers-in-law and sisters-in-law
Common people and employments
Relatives, one's in general
Foreign persons and losses through
Carpenters
Teeth, decayed or abscessed
Dental trouble
Enemies, secret, children
Children, in general
Foreign countries, people
Contractors
Gardeners, landscapers
Robbers, robberies
Deaths, sudden
Strangers
Unexpected, the
Assassination attempts
Neighbors, neighborhood
Execution, executioners
Enemies, secret, children of one's
Stooping, hunchbacks
Thin persons
Elderly persons
Friends of one's employees
Step families
Hate
Enemies, children of
Enemies, public and open
Illicit undertakings
Crimes of violence
Shooting, shootings
Domestic servants
Murder
Anger
Male relatives
Agitators
Bad breath, halitosis
Employees
Domestic servants
Servants, friends of one's
Housewives, housekeepers female household help

These words can then be categorized as follows:

<u>Possible link to information</u>:
Brothers and sisters, in general
Friends of one's brothers and sisters
Children, in general
Brothers-in-law and sisters-in-law
Relatives, one's in general
Step families
Male relatives

<u>Possible link to suspect areas</u>:
Foreign countries, people
Foreign persons and losses through
Enemies, children of
Enemies, public and open
Enemies, secret, children of one's
Strangers
Neighbors, neighborhood

<u>Possible link to suspect(s):</u>
Common people and employments
Servants, friends of one's
Housewives, housekeepers female household help
Domestic servants
Employees
Friends of one's employees

<u>Possible link to motive:</u>
Hate
Anger
Illicit undertakings
Robbers, robberies
Arguments with public enemies

<u>Possible link to outcome:</u>
Assassination attempts
Execution, executioners
Agitators

Possible link to crime type:
Crimes of violence
Shooting, shootings
Deaths, sudden
Unexpected, the
Murder

Possible link to employment suspect(s):
Common Employment
Contractors
Gardeners, landscapers
Carpenters

Possible link to suspect:
Neighbors, neighborhood
Foreign persons, and losses through
Bad breath, halitosis, dental trouble
Teeth, decayed or abscessed
Thin persons
Elderly persons
Stooping, hunchbacks

The Children

Martin, Jr., age 9, and Matthew, age 6, were the young victims in this murder case. Both boys had been shot with the same gun that killed their mother.

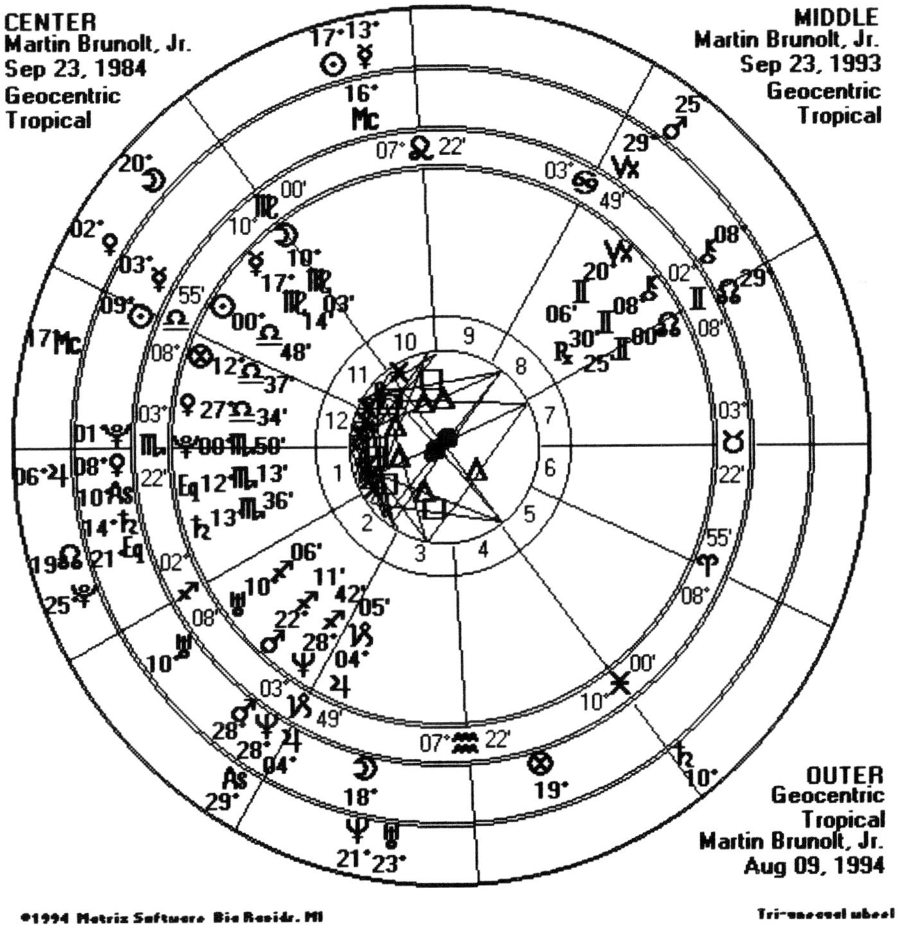

Chart 48 - Martin Jr.'s Three-ring Chart with Natal, Progressed Chart for 1993 and Transits Charts for August 9, 1994

Martin Brunolt, Jr.'s Natal Chart

Martin, Jr.'s natal chart is a bucket-type chart with all of the planets within 120 degrees of each other with the North Node acting as the handle at 00:25 Gemini R in the seventh house. The natal Sun is at the critical degree of 00:48 Libra in the eleventh house of hopes and wishes, and squares the South Node at 00:25 Sagittarius R in the first house of self. Natal Mercury is at the critical degree of 17:14 Virgo, also in the eleventh house. The closest aspect in his natal chart is the natal Moon at 10:03 Virgo in the eleventh house which squares Uranus at 10:06 Sagittarius in the second house.

There is a natal yod comprised of Neptune at 28:42 Sagittarius in the second house, North Node at 00:25 Gemini R in the seventh house, and Pluto at 00:50 Scorpio in the twelfth house.

The Progressed Chart for 1993

Martin, Jr.'s progressed East Point is at the critical degree of 21:06 Scorpio in the natal first house of self. The 29th degree is also prominent with the Nodes at 29:56 Taurus R and Scorpio, the Vertex at 29:03 Gemini in the natal eighth house of death and regeneration, and Neptune approaching the 29th degree at 28:48 Sagittarius in the natal second house.

There is a progressed yod comprised of progressed Mars at 28:14 Sagittarius and Neptune at 28:48 Sagittarius, both in the natal second house, the North Node at the ominous 29:56 degrees of Taurus R in the natal seventh house and Pluto at 01:10 Scorpio in the natal twelfth house.

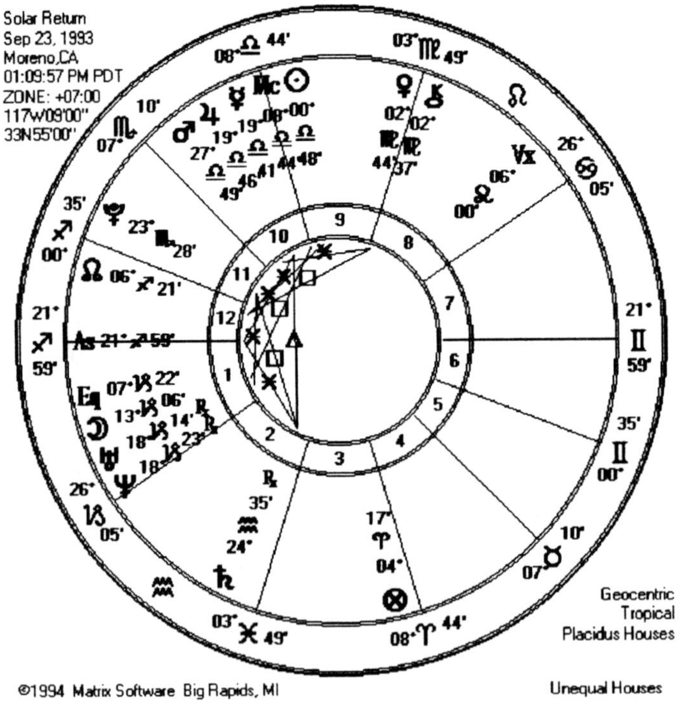

Chart 49 - Martin Brunolt, Jr.'s Solar Return Chart for 1993

The Solar Return Chart for 1993 (Moreno, CA)

The solar Moon is at the critical degree of 13:06 Capricorn in the solar first house of self and the solar Sun is at the critical degree of 00:48 Libra in the solar ninth house of the abstract mind. Neither the Moon or the Sun form any close aspects. The closest aspects in the solar return chart are the squares between Uranus at 18:14 Capricorn R and Neptune in conjunction at 18:23 Capricorn R, both in the solar first house of self, and Mercury at 19:41 Libra in conjunction with Jupiter at 19:46 Libra, both in the solar tenth house of public recognition.

An interception of Leo in the eighth house and Aquarius in the second house can indicate life and death matters during the year.

Matthew Brunolt's Natal Chart

In Matthew's natal chart Aries is intercepted in the first house of self and Libra in the seventh house of one-to-one relationships. A close T-square is formed by Mars at 29:44 Pisces in the first house of self, Mercury at 01:21 Cancer in the fourth house of the emotional and physical home opposing Midheaven at 29:43 Sagittarius, Saturn at 27:38 Sagittarius R, and Uranus at 28:09 Sagittarius R both in the house of abstract mind. Jupiter at 28:26 Taurus in the second house of finances and values also forms a close inconjunct to the Midheaven, Saturn and Uranus.

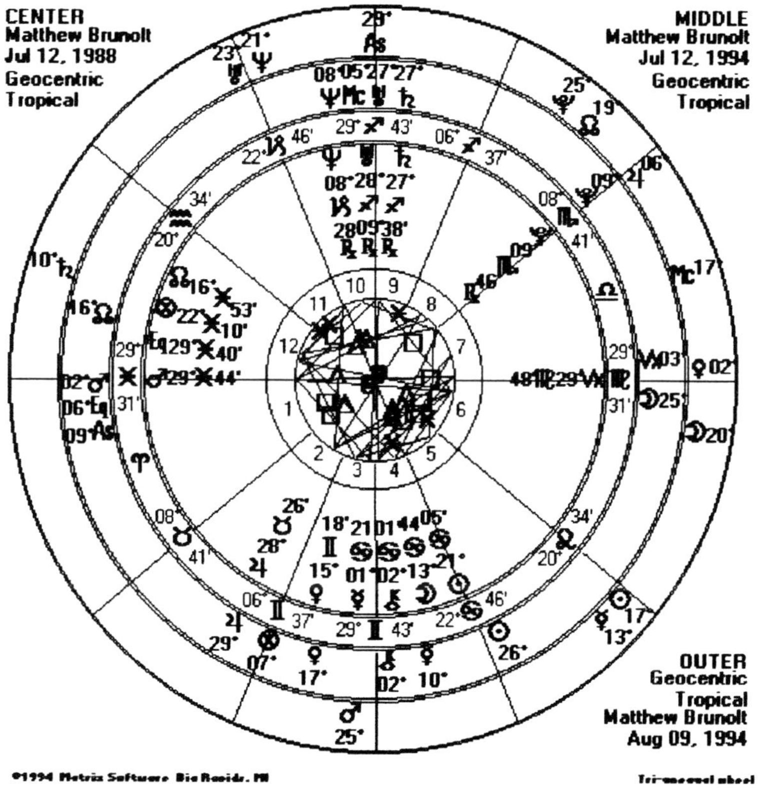

Chart 50 - Matthew Brunolt's Three-ring Chart with Natal, Progressed Chart for 1994 and Transits for August 9, 1994

The Progressed Chart for 1994

Progressed Sun at the critical degree of 26:48 Cancer in the natal fifth house of creativity and children sextiles the progressed Moon at 25:33 Virgo in the natal sixth house of health and service, and inconjuncts both progressed Saturn at 27:16 Sagittarius R and progressed Uranus 27:56 Sagittarius R both in the natal ninth house of abstract mind.

Progressed Pluto at the critical degree of 09:46 Scorpio R in the natal seventh house of others trines Mercury at 10:57 Cancer in the natal fourth house of the emotional and physical home sextiles Neptune at 08:19 Capricorn R in the natal tenth house of public recognition, and opposes the progressed Ascendant at 09:05 Aries.

Mercury at 10:57 Cancer forms a progressed T-square with Neptune at 08:19 Capricorn R and the Ascendant at 09:05 Aries. Progressed Jupiter at the ominous degree of 29:32 Taurus in the natal second house inconjuncts progressed Saturn at 27:16 Sagittarius R and Uranus at 27:56 Sagittarius R, both in the natal tenth house.

The Annular Solar Eclipse

The Annular Solar Eclipse on May 10, 1994 at 19:49 Taurus took place in Matthew's second house. A solar eclipse in the second house can indicate changes in one's financial situation and personal values.

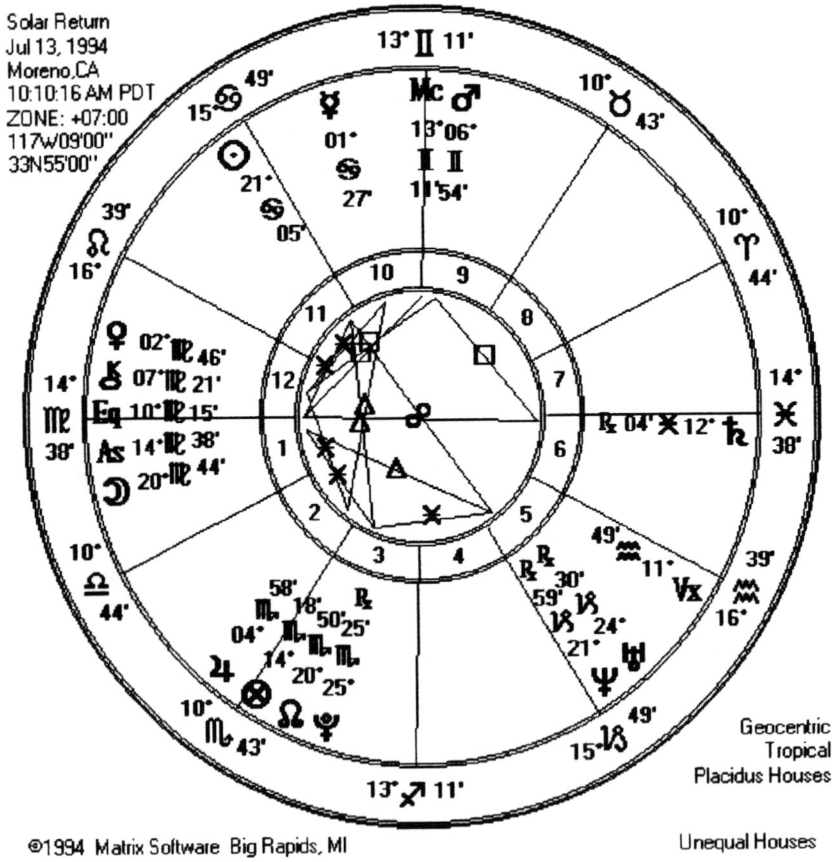

Chart 51 - Matthew Brunolt's Solar Return Chart for 1994

The Solar Return Chart for 1994 (Moreno, CA)

Solar Jupiter at 04:58 Scorpio in the second house of finances and values sextiles solar Venus at 02:46 Virgo in the twelfth house of secret enemies and inconjuncts solar Mars at 06:54 Gemini in the ninth house of abstract mind. Solar Saturn at 12:04 Pisces R in the sixth house of health and service to others forms a progressed t-square with the Midheaven at 13:11 Gemini and the Ascendant at 14:38 Virgo.

The Solar Sun at 21:05 Cancer in the eleventh house of hopes and wishes sextiles the solar Moon at 20:44 Virgo in the first house of self, sextiles the solar North Node at 20:50 Scorpio R in the third house of everyday environment, and opposes both solar Neptune at 21:59 Capricorn R and solar Uranus at 24:30 Capricorn R in the fifth house of children.

In Conclusion

Several years have passed since the shootings and Brunolt still has not gone to trial. The investigation into the deaths also raises questions about the evidence, which appears to be inconsistent, incomplete, illegible or even non-existent in some cases. Since then, several of the people who were involved with this case have come under investigation themselves. As the time has turned into years, will it be possible for Martin Brunolt to receive a fair trial?

Birth Data for Martin Lewis Brunolt:

Birth date:	May 23, 1949
Birth time:	08:30 PM/PDT
Birth place:	South Gate, CA
Longitude:	118W13 Latitude: 33N57
Source:	Birth certificate

Birth Data for Graciela Brunolt:

Birth date:	November 21, 1954
Birth time:	Unknown
Birth place:	Lourdes, El Salvador
Longitude:	Latitude:
Source:	Written by self, however, Lourdes is not included in the International Atlas or on Kepler for Windows.

Birth Data for Martin Lewis Brunolt:

Birth date:	September 23, 1984
Birth time:	09:20 AM/PDT
Birth place:	Anaheim, CA
Longitude:	117W55 Latitude: 33N50
Source:	Birth certificate

Birth Data for Matthew Henry Brunolt:

Birth date:	July 12, 1988
Birth time:	11:25 PM/PDT
Birth place:	Anaheim, CA
Longitude:	117W55 Latitude: 33N50
Source:	Birth certificate

Placements and degrees of Martin Brunolt's charts:
Natal Placements

Moon	20:26 Ari	04th	Neptune	12:38 Lib R	10th

Sun	02:40 Gem	05th	Pluto	14:19 Leo	08th
Mercury	17:10 Gem R	06th	North Node	23:52 Ari R	04th
Venus	12:30 Gem	06th	Vertex	07:47 Leo	08th
Mars	17:50 Tau	05th	East Point	09:52 Cap	01st
Jupiter	02:09 Aqu R	02nd	Fortuna	12:07 Sco	11th
Saturn	29:46 Leo	08th	Midheaven	11:40 Lib	
Uranus	29:02 Gem	07th	Ascendant	24:21 Sag	

Eleventh house position is 09:44 Sco
Twelfth house position is 02:56 Sag
Second house position is 28:56 Cap
Third house position is 06:58 Pis

Critical Degrees: Mercury
To be noted: Saturn at 29 degrees of Leo in the eighth house and Uranus at 29 degrees of Gemini in the seventh house
Intercepted Signs: Aquarius in the second house of finances and values and Leo in the eighth house of shared resources

Progressed Placements for May 1994

Moon	18:53 Sag	12th	Neptune	12:27 Lib	10th
Sun	15:40 Can	07th	Pluto	15:15 Leo	08th
Mercury	26:49 Gem	07th	North Node	21:29 Ari R	04th
Venus	07:34 Leo	08th	Vertex	05:06 Vir	08th
Mars	19:44 Gem	06th	East Point	19:54 Aqu	02nd
Jupiter	28:48 Cap R	01st	Fortuna	10:03 Can	07th
Saturn	02:59 Vir	08th	Midheaven	24:40 Sco	11th
Uranus	01:40 Can	07th	Ascendant	06:51 Aqu	02nd

Critical Degrees: None
To be noted: Jupiter retrograding from the 29th degree of Capricorn into the first house of self

Transiting Placements for August 9, 1994 at 3:45 PM/PST

Moon	20:58 Vir	09th	Uranus	23:27 Cap R	01st
Sun	17:09 Leo	08th	Neptune	21:17 Cap R	01st
Mercury	13:49 Leo	08th	Pluto	25:17 Sco	11th
Venus	02:30 Lib	09th	North Node	19:23 Sco R	11th
Mars	25:30 Gem	07th	Vertex	11:27 Leo	08th
Jupiter	06:56 Sco	10th	East Point	14:59 Cap	01st
Saturn	10:40 Pis R	03rd	Fortuna	03:11 Aqu	02nd

Critical Degrees: None
To be noted: None

Solar Return Placements for 1994 (Moreno, CA)

Moon	17:18 Sco	01st	Neptune	23:08 Cap R	03rd
Sun	02:40 Gem	07th	Pluto	26:31 Sco R	01st
Mercury	24:29 Gem	08th	No. Node	23:31 Sco R	01st
Venus	03:36 Can	08th	Vertex	05:39 Can	08th
Mars	00:05 Tau	06th	East Point	28:32 Sco	01st
Jupiter	06:57 Sco R	12th	Fortuna	01:36 Tau	06th
Saturn	11:40 Pis	04th	Midheaven	23:59 Leo	
Uranus	26:08 Cap R	03rd	Ascendant	16:58 Sco	

Eleventh house position is 26:19 Vir
Twelfth house position is 23:52 Lib
Second house position is 16:32 Sag
Third house position is 19:26 Cap

Critical Degrees: Uranus, the planet of sudden and unexpected events in the third house of everyday environment
To be noted: Mars at 00 degrees of Taurus in the sixth house of health
Intercepted Sign: None

Martin, Jr.'s Natal Placements

Moon	10:03 Vir	11th	Neptune	28:42 Sag	02nd
Sun	00:48 Lib	11th	Pluto	00:50 Sco	12th
Mercury	17:14 Vir	11th	North Node	00:25 Gem R	07th
Venus	27:34 Lib	12th	Vertex	20:06 Gem	08th
Mars	22:11 Sag	02nd	East Point	12:13 Sco	01st
Jupiter	04:05 Cap	03rd	Fortuna	12:37 Lib	12th
Saturn	13:36 Sco	01st	Midheaven	07:22 Leo	
Uranus	10:06 Sag	02nd	Ascendant	03:22 Sco	

Eleventh house position is 10:00 Vir
Twelfth house position is 08:55 Lib
Second house position is 02:08 Sag
Third house position is 03:49 Cap

Critical Degrees: Sun, Mercury
To be noted: The Sun at 00 degrees of Libra in the eleventh house of friends and acquaintances, Pluto at 00 degrees Scorpio in the twelfth house of secrets and psychic intunement, the North Node at 00 degrees of Gemini in the seventh house of one-to-one relationships
Intercepted Signs: None

Progressed Placements September 1993

Moon	18:41 Cap	03rd	Neptune	28:48 Sag	02nd
Sun	09:39 Lib	12th	Pluto	01:10 Vir	12th
Mercury	03:22 Lib	11th	North Node	29:56 Tau R	07th

Venus	08:35 Sco	01st	Vertex	29:03 Gem	08th
Mars	28:14 Sag	02nd	East Point	21:06 Sco	01st
Jupiter	04:52 Cap	03rd	Fortuna	19:45 Aqu	04th
Saturn	14:31 Sco	01st	Midheaven	16:13 Leo	10th
Uranus	10:24 Sag	02nd	Ascendant	10:43 Sco	01st

Critical Degrees: The East Point
To be noted: Neptune approaching the 29th degree of Sagittarius in the second house, North Node at the ominous sign of 29th degree of Taurus in the seventh house and Vertex at the 29th degree of Gemini in the eighth house

Transiting Placements for August 9, 1994

Moon	16:38 Vir	11th	Uranus	23:29 Cap R	03rd
Sun	16:51 Leo	10th	Neptune	21:19 Cap R	03rd
Mercury	13:11Leo	10th	Pluto	25:18 Sco	01st
Venus	02.10 Lib	11th	North Node	19:26 Sco R	01st
Mars	24:51 Gem	08th	Vertex	12:52 Pis	05th
Jupiter	06:50 Sco	01st	East Point	23:50 Vir	11th
Saturn	10:44 Pis R	05th	Fortuna	16:08 Lib	12th

Critical Degrees: None
To be noted: None

Solar Return Placements for 1993 (Moreno, CA)

Moon	13:06 Cap	01st	Neptune	18:23 Cap R	01st
Sun	00:48 Lib	09th	Pluto	23:28 Sco	11th
Mercury	19:41 Lib	10th	North Node	06:21 Sag R	12th
Venus	02:44 Vir	09th	Vertex	06:00 Leo	08th
Mars	27:49 Lib	10th	East Point	07:22 Cap	01st
Jupiter	19:46 Lib	10th	Fortuna	04:17 Ari	03rd
Saturn	24:35 Aqu R	02nd	Midheaven	08:44 Lib	
Uranus	18:14 Cap R	01st	Ascendant	21:59 Sag	

Eleventh house position is 07:10 Sco
Twelfth house position is 00:35 Sag
Second house position is 26:05 Cap
Third house position is 03:49 Pis

Critical Degrees: Moon, Sun
To be noted: None
Intercepted Sign: Leo intercepted in the eighth house and Aquarius intercepted in the second house

Matthew's Natal Placements

Moon	13:44 Can	04th	Neptune	08:28 Cap R	10th
Sun	21:05 Can	04th	Pluto	09:46 Sco R	08th
Mercury	01:21 Can	04th	North Node	16:53 Pis R	12th
Venus	15:18 Gem	03rd	Vertex	29:48 Vir	07th
Mars	29:44 Pis	01st	East Point	29:40 Pis	01st
Jupiter	28:26 Tau	02nd	Fortuna	22:10 Pis	12th
Saturn	27:38 Sag R	09th	Midheaven	29:43 Sag	
Uranus	28:09 Sag R	09th	Ascendant	29:31 Pis	

Eleventh house position is 22:46 Cap
Twelfth house position is 20:34 Aqu
Second house position is 08:41 Tau
Third house position is 06:37 Gem

Critical Degrees: Moon
To be noted: Mars at 29 degrees, Vertex at 29 degrees, East Point at 29 degrees, Midheaven at 29 degrees, and the Ascendant at 29 degrees
Intercepted Signs: Aries intercepted in the first house and Libra intercepted in the seventh house

Progressed Placements for July 1994

Moon	25:35 Vir	06th	Neptune	08:19 Cap R	10th
Sun	26:48 Can	05th	Pluto	09:46 Sco R	08th
Mercury	10:57 Can	04th	North Node	16:34 Pis R	12th
Venus	17:39 Gem	03rd	Vertex	03:55 Lib	07th
Mars	02:24 Ari	01st	East Point	06:27 Ari	01st
Jupiter	29:32 Tau	02nd	Fortuna	07:52 Gem	03rd
Saturn	27:16 Sag R	09th	Midheaven	05:26 Cap	
Uranus	27:56 Sag R	09th	Ascendant	09:05 Ari	

Critical Degrees: Sun, Pluto
To be noted: Jupiter at 29 degrees

Transiting Placements for August 9, 1994

Moon	20:58 Virgo	06th	Uranus	23:27 Cap R	11th
Sun	17:09 Leo	05th	Neptune	21:17 Cap R	10th
Mercury	13:49 Leo	05th	Pluto	25:17 Sco	08th
Venus	02:47 Lib	07th	North Node	19:23 Sco R	08th
Mars	25:30 Gem	03rd	Vertex	11:44 Leo	05th
Jupiter	06:56Sco	07th	East Point	15:16 Cap	10th
Saturn	10:40 Pis R	12th	Fortuna	03:32 Aqu	11th

Critical Degrees: None
To be noted: None

Solar Return Placements for 1994 (Moreno, CA)

Moon	20:44 Vir	01st	Neptune	21:59 Cap R	05th
Sun	21:05 Can	11th	Pluto	25:25 Sco R	03rd
Mercury	01:27 Can	10th	North Node	20:50 Sco R	03rd
Venus	02:46 Vir	12th	Vertex	11:49 Aqu	05th
Mars	06:54 Gem	09th	East Point	10:15 Vir	12th
Jupiter	04:58 Vir	02nd	Fortuna	14:18 Vir	03rd
Saturn	12:04 Pis R	06th	Midheaven	13:11 Gem	
Uranus	24:30 Cap R	05th	Ascendant	14:38 Vir	

Eleventh house position is 15:49 Can
Twelfth house position is 16:39 Leo

Second house position is 10:44 Lib
Third house position is 10:43 Sco

Critical Degrees: Jupiter
To be noted: None
Intercepted Sign: None

THE CON-ARTIST

Sherry Diamond alias Sheri Kaye
The Shrimp Lady

On April 26, 1989, the West Allis, Wisconsin police arrested Sherry Diamond alias Sheri Kaye and found $248,000 in cash in the trunk and back-seat of her 1988 Cadillac. When the police questioned her about the money, she told them that her mother, who traveled with her, had inherited it from a relative. The *Schenectady Daily Gazette* reported that police also found that Diamond traveled with an extensive wardrobe. It consisted in part of 330 blouses, 709 pieces of lingerie, 276 scarves, 295 pairs of earrings, 160 skirts, 104 sweaters, five fur coats, and 572 pieces of jewelry.

Saturday, October 9, 1993 — THE DAILY GAZETTE

'Shrimp Lady' fires lawyer after dispute

By TIM McGLONE
Gazette Reporter

SCHENECTADY — Sherry Kaye, the so-called Shrimp Lady, fired her attorney Friday after she rejected a plea bargain that would have set her free with about $80,000 of the $404,000 confiscated by police.

Kaye, 42, spent Thursday and part of Friday in County Court negotiating a plea bargain with the district attorney's office, prosecutors said.

A deal was worked out that called for her to plead guilty to a felony charge and be sentenced to six months in jail — time she has already served — and face fines and forfeitures of about $45,000.

"In the past few weeks, there has been a great deal of contention. Our conversations broke down," Kaye told visiting Judge Gene L. Catena of her relationship with attorney Brian Premo.

"I would not say they were contentious. They were anything but," Premo told the judge.

Kaye argued that Premo has been "desperate to withdraw" himself from the case, but Premo denied that and said Kaye already fired him earlier in the week.

Kaye faces an eight-count indictment charging her with scheming to defraud six Price Chopper supermarkets in the Capital Region between May last year and April this year.

In each case, she allegedly returned food — usually frozen shrimp — for a cash refund, though she never purchased the food to begin with.

She allegedly stole from stores in Glenville, Wilton, Rotterdam, Schenectady, Colonie and Guilderland.

She pleaded guilty in Wilton Town Court in July to second-degree scheme to defraud, a misdemeanor. She admitted that on May 22 last year she collected a $71.91 refund for frozen shrimp she never purchased.

with Premo. In another, a judge disagreed.

Premo said in court Friday that although he believes Kaye would face double jeopardy in Schenectady, he advised Kaye that a County Court judge may not agree.

With that, negotiations for a plea bargain were started. Premo said in court that Kaye agreed to the plea bargain on Wednesday.

"She then advised my office yesterday that she did not want to take it," Premo said in court. She then fired him, he said.

But Kaye returned to him on Friday and asked him to represent her in court for the day.

"Do you want to discharge him?" Catena asked Kaye in court Friday.

"I'm going to let him go. I just don't feel he is very happy with this arrangement," Kaye said.

Catena then released Premo and he walked out of court.

Kaye then tried to act as her own attorney and sought to speak privately with Chief Assistant District Attorney Alfred D. Chapleau.

"I don't want to talk to her," Chapleau told the judge. Chapleau then rescinded the plea bargain offer.

The plea bargain called for Kaye to plead guilty to first-degree scheme to defraud, a felony.

She would have been sentenced to six months in jail and fined $5,000. She also would have had to pay $1,000 restitution to Price Chopper as well as $39,000 to the county as a forfeiture for the book value of the 1992 Cadillac that police seized from her when she was arrested.

The Internal Revenue Service has already seized $276,288 for payment of taxes owed on the $404,000 confiscated by police, leaving a balance of about $128,000.

If she would have accepted the plea bargain, she would have been left with about $80,000 plus the Cadillac

MARC SCHULTZ Gazette Photographer
Sherry Kaye, the so-called "Shrimp Lady," walks into Schenectady County Court Friday with her attorney, Brian Premo. She later ...

Illustration 28 - The Daily Gazette (Sherry Diamond, alias Sherry Kaye)

This mother and daughter had conned their way around the country, getting free meals and motel rooms. Their grocery scam seldom varied. Diamond usually entered a market and picked up bags of frozen shrimp or occasionally prime cuts of beef and chicken. She then took them to the service counter and said she had bought them earlier but they were rotten, and asks for a refund.

Since Diamond's arrest in 1989, the mother and daughter team had rebuilt their nest egg. When Glenville police arrested Diamond on April 9, 1993 in Upstate New York, they seized her 1992 Cadillac, which had $408,000 in the trunk. The police investigation found that she had paid cash for the car in Oregon. It was the second car Diamond had paid cash for in four years. Police also seized a small travel trailer Kaye pulled behind her car; it was filled with clothes, jewelry, and other belongings which they have valued at $100,000. The longitude and latitude for Glenville, NY are 074W03 and 42N55.

An alert store clerk suspected that Diamond was lying when she called the market and said she was bringing back some shrimp for a refund. The clerk called the police and marked all the packages of shrimp in the store with a black magic marker. Diamond was arrested when she returned six packages of shrimp all having a black mark on them. Police did not charge Diamond's mother, Diana Kaye, who is in her late 70s, with any crime.

Diamond spent almost five months in the Upstate New York jail in lieu of $100,000 in bail after her arrest on April 9th for a series of grocery store scams. The indictment charged Diamond with four counts of scheme to defraud, which is a felony, three counts of petty larceny, and one count of attempted petty larceny, which are misdemeanors. The indictments stemmed from charges in Albany, Saratoga, and Schenectady Counties.

The August 20, 1993 headlines of the *Schenectady Daily Gazette* read, "'Shrimp Lady' released from jail after ruling." The Glenville Police still had control of the $408,000, the 1992 Cadillac and the trailer of belongings they seized in April when Diamond was arrested. However, they feared that the mother and daughter team, who left behind $125,000 in bail money in Wisconsin after Diamond's release from jail in 1989, might flee the area. The IRS had also placed a lien on the money for $262,000 in unpaid taxes owed on the $408,000.

Sherry Diamond, alias Sheri Kaye, was born in San Diego, California. Her birth certificate lists Diana Kaye and Murray Diamond as her mother and father, both of whom were born in New York State. Records show she has been arrested about twenty times in ten states since 1971. In addition, the 'Shrimp Lady' has several aliases, five driver's licenses, and many birth certificates.

Diamond's Natal Chart

Looking at Diamond's natal chart, we see a T-square. The T-square involves the Moon at 12:03 Libra in the eighth house of other's possessions, Mercury at the critical degree of 13:48 Capricorn in the twelfth house of self-undoing and confinement and Uranus at 15:06 Cancer in the sixth house of work. Uranus in Cancer in the sixth house suggests a restless cleverness in her ideas of employment. There is also a need to work in a profession that allows her the freedom to move around.

Diamond's natal Aquarian Sun at 07:52 Aquarius and Venus at 25:57 Aquarius in the first house of self allows her success in presenting herself uniquely within her environment. Natal Mars at the critical degree of 04:46 Pisces, also in the first house, gives Diamond an ability to act and think quickly in confronting situations.

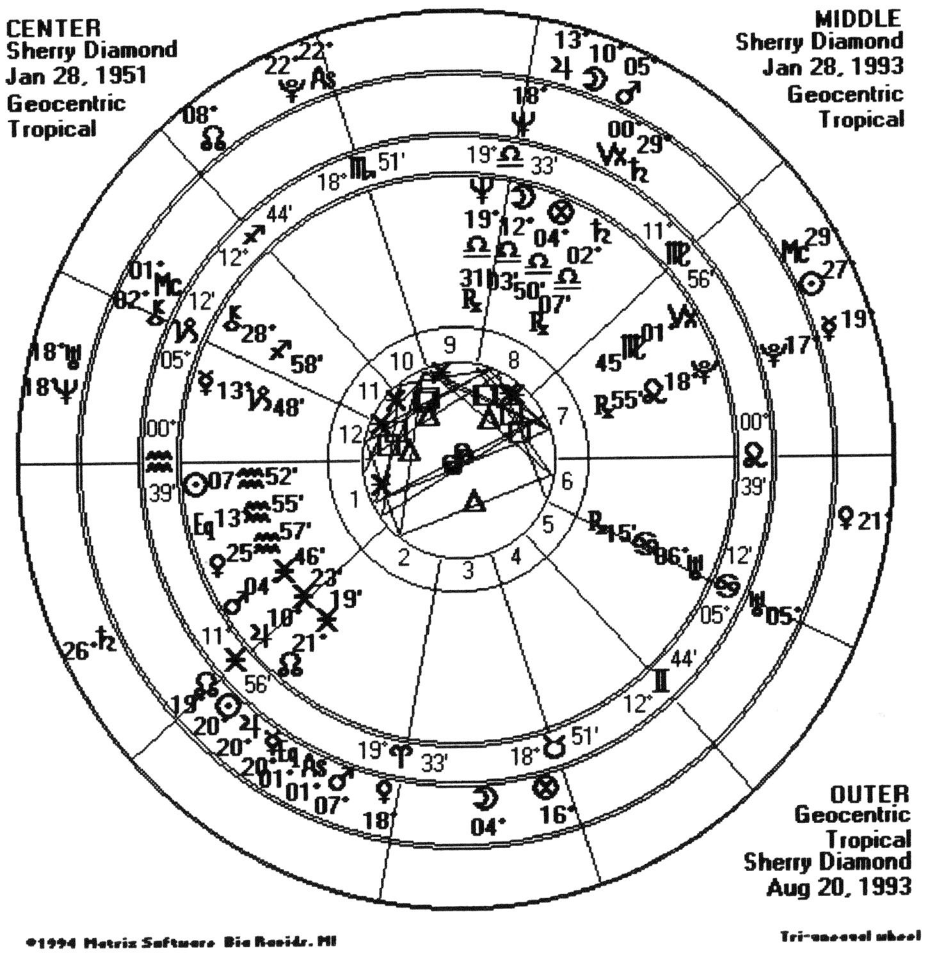

Chart 52 - Sherry Diamond's Three-ring Chart with Natal, Progressed Chart for 1993 and Transits for October 22, 1993

Saturn at 02:07 R Libra in the eighth house gives Diamond the Librans' intelligence and an understanding of the law and its workings. It is also indicative of her lessons to be learned in connection with the money and possessions of others. The Moon at 12:03 Libra and Neptune at 19:31 Libra R are also in the eighth house. The Moon suggests an easy-come, easy-go attitude regarding these shared resources. At the same time, Diamond's attitude can be easily influenced by her mother. Neptune in Libra warns of her deceptive charm. It should be remembered that Libra can be either the artist or as we see in Sherry Diamond's chart, the con-artist. Her Mercury is placed at the critical degree of 13:48 Capricorn in the twelfth house of self-undoing which forewarns of confinement or imprisonment.

Diamond's Progressed Chart for 1993

Of interest in Diamond's progressed chart for January 28, 1993 is the close conjunction of progressed Sun at 20:13 Pisces, Jupiter at 20:18 Pisces and Mercury at 20:23 Pisces in the natal second house of personal values and finances. Pisces is also the ruler of the twelfth house of self-undoing and imprisonment.

Progressed Uranus has recently ingressed into the natal sixth house, while progressed Saturn is at the "turning point" degree of 29:41 Virgo in the natal eighth house of other's possessions and values. Additionally, the progressed Vertex is at the critical degree of 00:51 Libra in the eighth house.

Transiting Uranus-Neptune

The transiting Uranus-Neptune conjunction of 1993 is found in Diamond's twelfth house of self-undoing and imprisonment. It is reported that police first began their investigation of Diamond in 1992. It can be assumed that they had gathered much of their evidence and were preparing the warrant for her arrest with the first conjunction of these two planets in early February 1993. As stated earlier, Diamond was arrested in April and spent the next five months in jail. On August 20, 1993 and the second Uranus-Neptune conjunction, she was released from jail. The third and final conjunction of Uranus-Neptune took place between October 22 and October 27, 1993.

On October 22, Diamond went to the Price Chopper Market in Schenectady, New York and claimed she had paid good money for some bad oranges. The manager put a "paid" sticker on another bag and told Diamond to go and replace the oranges. Security watched as she took the oranges, but she also decided to take some grapes, muffins, croissants and then walked out the door. Police arrested her and she was again sent to jail for petty larceny in connection with taking $19.25 worth of groceries. Her mother posted her daughter's bail of $10,000 in small bills for her release on the same day.

A Partial Solar Eclipse on May 21, 1993 occurred at 00:32 Gemini. The eclipse occurred in Diamond's fourth house, which indicates forced changes in one's home environment, moving from one home to another, or responsibility concerning a parent.

In Conclusion

On October 29, 1993, Sherry Diamond, the so-called "Shrimp Lady," was sentenced to pay $45,000 in fines, forfeitures and restitution to several supermarkets. Diamond is entitled to get back her 1992 Cadillac and police are still holding more than $126,000 of the seized money. The authorities will take $39,000 as a forfeiture for the value of the car, various fines and restitution, after which there will be about $80,000 remaining. Half of this money will be paid to her attorney, leaving Diamond with about $40,000. However, she did not attempt to retrieve her money, car or the trailer full of clothes.

There is an open petty larceny charge against Diamond from the October 22, 1993 incident. However, an employee of the local motel where she and her mother were staying advised authorities that the two had checked out on October 29th. No one has seen Sherry Diamond in upstate New York, but she may be in another part of the country, rebuilding her nest egg and conning the unsuspecting.

In June 1995, the Glenville police department took delivery of a shipment of .40 caliber semiautomatic pistols which were paid for by the Town's share of the 'Shrimp Lady's' confiscated loot.

In Glenville, a hello to arms

Shrimp Lady's seized booty does police a jumbo favor

By Joe Mahoney 06/04/95
Staff writer

GLENVILLE — Thanks to the Shrimp Lady, town police say they will soon be obtaining powerful new service weapons.

Glenville Police Chief Jack Purdy said his department is expected to take delivery this month of a shipment of .40-caliber semiautomatic pistols, giving Glenville officers greater stopping power in the event they find themselves in a gunbattle.

And the money is coming out of the town's share of the loot confiscated from Sherry Kaye, who became known as the Shrimp Lady in April 1993 after she was arrested in a massive shoplifting enterprise.

Kaye, who eventually pleaded guilty to scheming to defraud, bilked several local supermarkets by stealing hefty bags of seafood and then returning them to the markets for cash exchanges. When Glenville police stopped the scam, they searched Kaye's late-model Cadillac and seized $404,000 in cash. Most of the money went to the Internal Revenue Service.

Purdy said his 19 officers will be shedding their 9mm handguns when the new Ruger semiautomatics arrive. The Rugers cost about $300 each, he said.

"It is a higher-caliber round, and we tested it on different types of materials," he said.

The bottom line, he said, is that it gives police officers a better chance of surviving a gunbattle because, compared with the 9mm semiautomatics, it fires a deadlier round.

Part of the impetus for moving to the .40-caliber weapon was a July 1993 encounter outside the Scotia Motel between Glenville Police Officer Richard D. Conley and John Orzolek, who police say pointed a hunting rifle at the officer after firing a shot in the direction of motel residents. Police said the officer then aimed his 9mm handgun at Orzolek and fired two shots, with one of the bullets ripping into the rifleman's cheek and exiting through his neck.

Even after being shot, Purdy said, "the individual was still capable of fighting with the officer. Our officer had to wrestle him to the ground. (Orzolek) was still combative when they got him to the hospital."

The only disadvantage to the .40-caliber is that it holds 11 rounds in its clip, four fewer than the 9mm, he noted. "But in the average police shooting, less than three rounds are fired by the officer," he added.

Most police agencies in the Capital Region are equipped with 9mm guns now, moving to the semiautomatics from what had been standard fare .357-caliber revolvers.

Illustration 29 - Albany Times Union (Diamond)

Birth Data for Sherry Diamond

Birth date:	January 28, 1951		
Birth time:	06:26 PM/PST		
Birth place:	San Diego, CA		
Longitude:	117W09	Latitude:	32N43
Source:	Birth certificate		

Placements for Sherry Diamond's charts:
Natal Placements

Moon	12:03 Lib	08th	Neptune	19:31 Lib R	08th
Sun	07:52 Aqu	01st	Pluto	18:55 Leo R	07th
Mercury	13:48 Cap	12th	North Node	21:19 Pis R	02nd
Venus	25:57 Aqu	01st	Vertex	01:45 Vir	07th
Mars	04:46 Pis	01st	East Point	13:55 Aqu	01st
Jupiter	10:23 Pis	01st	Fortuna	04:50 Lib	08th
Saturn	02:07 Lib R	08th	Midheaven	18:51 Sco	
Uranus	15:06 Can R	06th	Ascendant	00:39 Aqu	

Eleventh house position is 12:44 Sag
Twelfth house position is 05:12 Cap
Second house position is 11:56 Pis
Third house position is 19:33 Ari

Critical Degrees: Mercury and Mars
To be noted: Ascendant at 00 degrees
Intercepted Signs: None

Progressed Placements for 1993

Moon	04:37 Tau	03rd	Neptune	18:54 Lib	08th
Sun	20:13 Pis	02nd	Pluto	17:57 Leo	07th
Mercury	20:23 Pis	02nd	North Node	19:06 Pis R	02nd
Venus	18:01 Ari	02nd	Vertex	00:51 Lib	08th
Mars	07:28 Ari	02nd	East Point	01:26 Ari	02nd
Jupiter	20:18 Pis	02nd	Fortuna	16:23 Tau	03rd
Saturn	29:41 Vir	08th	Midheaven	01:12 Cap	11th
Uranus	05:24 Can	06th	Ascendant	01:59 Ari	02nd

Critical Degrees: Vertex in the eighth house of other's resources
To be noted: Saturn at 29 degrees in the eighth house of other's resources

Transiting Placements for August 20, 1993

Moon	12:10 Sco	08th	Uranus	18:47 Cap R	12th
Sun	27:48 Leo	07th	Neptune	18:48 Cap R	12th
Mercury	05:19 Leo	07th	Pluto	22:49 Sco	10th
Venus	21:56 Can	06th	North Node	08:09 Sag R	10th
Mars	05:34 Lib	08th	Vertex	12:04 Can	06th
Jupiter	03:13 Lib	08th	East Point	04:00 Sag	10th
Saturn	26:53 Aqu R	01st	Fortuna	04:33 Cap	11th

Critical Degrees: East Point
To be noted: None

Your comments or questions are welcome on any of the cases in this book. Write to:

Arlene L. DeAngelus
18 Montgomery Way
Clifton Park, NY 12065

E-mail: ADeAngelus@AOL.com

WORKS CITED

FBI, Uniform Crime Reports, *1998 Preliminary Annual Release*, http:/www.fbi.gov/ucr/prelim98.pdf
The Washington Post, August 25, 1997
James Tedesco and Michele Paludi, *Missing Children*
Ivy M. Goldstein Jacobson, *Simplified Horary Astrology*
Vivian E. Robson, *The Fixed Stars & Constellations in Astrology*
Rex E. Bills *The Rulership Book*
Lois R. Rodden, *Astro-Data III*
Lois M. Rodden, *Astro-Data V*
Robert Ebertin, *Transits*
Frances Sakoian and Louis Acker, *Predictive Astrology*
Lois Rodden, *Modern Transits*
Sophia Mason, *Forecasting with the New, Full & Quarter Moons Through the Houses*
Michael Munkasey, *Astrological Thesaurus*
Lois Rodden, *Data News* #46 April 1994

Astrological Crime Charts is not a book of interpretations, but instead it is a "What To" and "How To" of interpreting Crime Charts. *Crime Charts* gives you the background of the victim as well as the astrological charts for each case. In addition to aspects and house placements, yods, critical degrees, the 29 and 00 degrees of a sign and eclipses are noted for your easy reference. Learn to find additional clues pertaining to the crime in the victim's astrological charts.

PHOTO TO BE SUPPLIED

Arlene DeAngelus has been involved in Astrology for over 25 years and lectures nationally on Crime Charts. She has a BA in Communication Arts and attends graduate classes at The College of Saint Rose in Albany, New York. She is a member of the American Federation of Astrologers, the National Council for GeoCosmic Research and The Astrological Association of Great Britain.